Politics, Propaganda, and Pop Culture

Politics, Propaganda, and Pop Culture

Cold War Television Cultures

Edited by

Christopher Porter, Anthony Salisbury, and Tyler Ripley

Executive editor:

Heather Gumbert

Chapter authors:

Aaidin Finefield; Anthony Salisbury; Cathryn Jones; Chase Frazier; Christopher Porter; Daniel Smith; Danielle Schumacher; Jason Toy; Jeremiah Albert; Joshua Buckingham; Stratis Bohle; and Tyler Ripley

LIBERAL ARTS AND HUMAN SCIENCES HISTORY VIRGINIA TECH

VIRGINIA TECH. PUBLISHING

Virginia Tech Department of History in association with Virginia Tech Publishing

Blacksburg, VA

First published in 2024 by Virginia Tech Department of History in association with Virginia Tech Publishing

Virginia Tech Department of History
431 Major Williams Hall
220 Stanger Street
Blacksburg, VA 24061

Virginia Tech Publishing
University Libraries at Virginia Tech
560 Drillfield Drive
Blacksburg, VA 24061

Cataloging-in-Publication Data
Politics, Propaganda, and Pop Culture / edited by Christopher Porter, Anthony Salisbury, and Tyler Ripley, with Heather Gumbert and the Virginia Tech spring 2023 students of HIST 4914. Blacksburg, Virginia: Virginia Tech Department of History in association with Virginia Tech Publishing, 2024. Includes bibliographic references. This collection of essays began as a class project for a Virginia Tech Department of History undergraduate capstone research seminar (HIST 4914), examining the role of television in projecting, shaping, and amplifying Cold War ideologies.

ISBN (print): 978-1-962841-06-1 | ISBN (PDF): 978-1-962841-08-5 | ISBN (EPUB): 978-1-962841-07-8
DOI: https://doi.org/10.21061/coldwartelevisioncultures

History–Modern–20th Century–Cold War. 2. Performing Arts–Television–History & Criticism. 3. Social Science–Media Studies. 4. History–Wars & Conflicts–General. I. Gumbert, Heather

Cover
Image: https://commons.wikimedia.org/wiki/File:A_Colour_Television_Test.jpg (CC BY SA)
Design: Stratis Bohle

Contents

Dedication

To Dr. Gumbert and all of the other Virginia Tech history professors, without whom we never would have gotten to where we are.

Part I: Politics

The "politics" category of the book discusses the many roles television played in maintaining establishing new ideas of—as well as exercising control over—nation and ideology. Television played a critical role in state-building during the reconstruction era after the end of World War II, both by offering new uniting images and by projecting ideals onto the people. Television shaped perceptions of the world for audiences, influencing them with the ideals of those in charge of production. Governments latched onto this in different ways, either by creating their own stations to control the output of the medium, or by televising influential events to create a form of connection with the audience.

Crowning a Commonwealth

STRATIS BOHLE

If it was not televised, did it truly happen? The live broadcast of Queen Elizabeth II's Coronation in 1953 was a significant moment in the history of television broadcasting. It was the first time that a British coronation and major event was broadcast live or as close to live as possible across the globe. It marked a turning point in the way that people perceived television as a medium of communication. Television was only twenty years old and was deemed useful for entertainment purposes but did not yet have the same respect that radio had earned during World War II. This was due to its high cost to consumers before the war and the fact that television service in Great Britain was suspended during WWII. Furthermore, the rise of television catapulted the British monarchy into the homes of every Briton, Canadian, Australian, and all other commonwealth nations as the people's sovereign. It made the monarchy, which had been so distant suddenly come close to them, allowing the people to view the Queen, as schoolchildren look upon fish in a fishbowl. The hundreds of years of disseminating and controlling information by the British monarchy were slowly disappearing. The question, "Did it really happen?" suddenly gave new meaning to the news. The adaptation by the British monarchy in its role and actions throughout the empire served to perpetuate itself even at the loss of tradition and traditional communication mediums. Television made events seem more real to the people.

The telecast of the Coronation of Queen Elizabeth II marked a profound shift in sentiment regarding the significance of television. The Coronation of King George VI (her father) had not been televised due to the belief that it was inappropriate to publicly display a sacred event on such a rudimentary device.[1] By the early 1950s, however, the decision to transmit the events of the Coronation marked a new belief that television could be utilized for the purposes of soft diplomacy by exporting an event worldwide. The British monarchy, once reluctant to embrace television, adapted to survive, as it always had. Millions of people worldwide watched the Coronation. By showcasing the British monarchy and its role within the Commonwealth, the new medium of television also demonstrated its own power to define and project a new vision of the British Empire, reframing the relationship between the Commonwealth countries and postwar Britain.

GOD, KING, AND EMPIRE

Beginning with the Victorian era, Britain experienced significant prosperity due in large part to the expansion of its overseas empire, most notably in what became British India and the British African colonies. The idea of a British Empire began in the Age of Exploration but did not become what we know today until the nineteenth century. To understand how Britain and its monarchy adapted through developments within the Empire and their interactions in the past 150 years we have to look at three major events: Queen Victoria becoming Empress of India, furor over the Germanic ties of the Royal Family, and the Abdication Crisis of Edward VIII.

As part of the expansion into India and Africa, the holdings of British protectorates or colonies had vastly increased since the ascension of Queen Victoria in 1836. With the dissolution in 1858 of the East India Company (EIC), which had ruled British India, Britain began direct rule and affirmed the deposition of the Mughal Emperor, the ruler of India for the previous 300 years, by the EIC the year before.[2] The idea of proclaiming the British

Empire was proposed before direct rule, with the earliest occurring in the 1840s. In 1877, the title Empress of India was bestowed upon Queen Victoria, marking the *de jure* beginning of the British Empire.[3] In addition to the previous attempts at creating an imperial title, Russia and Austria-Hungary were no longer the only European imperial states after the creation of the German Empire. The Princess Royal, Victoria, was married to Frederick William, Crown Prince of the German Empire, and would soon outrank her mother, Queen Victoria, upon her husband's ascension to the throne. These events culminated in the creation of the title Emperor of India, which was maintained until after the partition of British India in 1947.

Another factor that affected the British Royal Family and their role in the Empire was the anti-German sentiment during the First World War. A number of Queen Victoria's children were either married to Imperial German Royals or, in the case of her son Alfred, had become ruling members of the German Empire.[4] These children maintained their British titles, and if they held British peerages stood in the House of Lords, the upper house of the British Parliament. But once the First World War began, the monarchy's German connections became a liability. The British government began a propaganda movement against Germany, using cartoons and posters to demonize the German foe and inadvertently bringing attention to the House of Saxe-Coburg and Gotha.[5] King George V took austere measures to address the controversy including changing the name of the Royal House to the more anglicized name Windsor.[6] In addition, the Titles Deprivation Act 1917 stripped all British titles from those who held allegiance to Germany, as it was a concern that they may return to legislate after the war.[7] In doing so, the King adapted the modern Royal Family to a changing Britain by minimizing the Germanness that had been an integral part of the monarchy since 1714 with the ascension of the first of the Hanoverian kings.

The third major event was King Edward VIII's abdication crisis due to his relationship with Wallis Simpson. The King's romantic relationship with the married but soon-to-be-divorced American socialite sparked concern within the more conservative circles of the British government and Empire who objected to the King's actions due to his role as head of the Church of England. Rather than give up his relationship with Simpson, King Edward VIII abdicated the throne less than a year into his reign and before his actual coronation. His brother the Duke of York succeeded him and ascended the throne as King George VI. The crisis came at a difficult time for the monarchy, due to a small but growing republican movement. By abdicating, King Edward VIII was able to secure the monarchy's future and showed that the monarchy could survive scandal by passing the throne to an heir.[8]

THE WAR

King George VI ascended the throne when the country was still recovering from the Great Depression and helped lead it with Prime Minister Winston Churchill through the Second World War. Just three years into his reign the war broke out; the King and his immediate family refused to leave London, which made them quite popular among the people. The young Princesses had the opportunity to be sent to Canada but stayed behind showing the English people and the world their dedication and sense of duty to their country.[9] The Princesses Elizabeth and Margaret embraced their duties, speaking directly to all the children that had fled London during the war through a program on the on the BBC Radio Service called the *Children's Hour*.[10] King George VI and his family helped to lead the British and stiffen Imperial resolve in fighting the German and Japanese armed forces. In doing so the King and the Royal Family became a popular cultural symbol within the nation and empire.

Ever since the Glorious Revolution of the seventeenth century, the British monarchy has not held much of a political role beyond the symbolic control of the country through the sovereign's government. Since then, the monarchy has shifted to being a cultural touchstone for Britain and the Empire. As part of this role, the monarchy embodied the values of duty, service, and sacrifice. In embodying these characteristics members of the Royal Family tended to join the armed forces and during the Second World

War, this was no exception. King George VI's brother the Duke of Kent was killed during a Royal Air Force Flight, and the King's daughter, Princess Elizabeth served in the Auxiliary Territorial Service once she turned 18 in 1944.[11] Even though television would not resume broadcasting until 1946, the Princess' actions were made public through the media available including radio and newspaper publications.

Serving her country as the war wound down helped the future Queen gain a reputation as a dedicated member of the Royal Family. Queen Elizabeth II was already popular by the time of her ascension on February 6, 1952, following her father's passing. Part of this popularity was due to her wedding in 1947 to Philip Mountbatten, a minor Greek royal who had lived in exile in the UK since the second dissolution of the Greek monarchy in 1921.[12] Mountbatten, who was distantly related to Elizabeth, served his adoptive home in the Royal Navy during the Second World War. Prince Philip's wedding to Elizabeth was the first major royal event since the war's conclusion, and it helped mark the recovery from the destruction by the Blitz and the destruction of London. The press followed the preparations closely. Due to the war's destruction, the rationing stamp system was still in place in 1947; press reports noted approvingly that the Princess used the rationing stamp system to purchase the goods necessary for her wedding dress.[13] This attempt to maintain the struggle of the common Briton for such a lavish event for which there would be state dignitaries was extremely well received within the country and helped the public to perceive Princess Elizabeth as "one of the people." The wedding was also the first major royal event covered by the BBC television service alongside the BBC radio service, though as it was performed at Westminster Abbey, the television service could not televise the religious service itself. The utilization of the press and the BBC helped refashion the relationship between the royal family and the British people.

In the aftermath of the Second World War, the process of decolonization that had begun with granting Dominion status to Canada, Australia, New Zealand, and South Africa accelerated. Britain had grown weaker, due to wartime damage inflicted upon the United Kingdom proper including debts incurred, a significant loss of life, and the need to rebuild the country. More important, resistance to European colonial rulers had stiffened during World War II. The capacity to maintain control over the British colonies was not there, while colonial subjects increasingly pressed for independence, and something had to give.

The Dominion system created what was originally called the British Commonwealth of Nations. British Parliament granted the status of Dominion to the previously mentioned states with the Balfour Declaration of 1926. The Declaration mandated that Dominion states were autonomous with no hierarchy and no mother country, though all unified through the Crown, and as such would be members of the Commonwealth of Nations. It was through this Balfour Declaration, and subsequent Statute of Westminster 1931, that the term Commonwealth of Nations became an official term to describe the self-governing members of the British Empire.[14]

In British India, the movement for self-rule had been moving along steadily before the war, with the prior decades characterized by widespread anti-colonial sentiments and the desire for autonomy and independence from the United Kingdom. In 1947 British India was given the ability to rule in the same manner as the countries listed prior with the granting of Dominion status. The only difference was that these previously mentioned Dominions were formed through the confederation or union of previously separate colonies, while British India was split in two by granting Dominion status to the new states of India and Pakistan. Britain could not turn back on the process of decolonization. Even though television had no influence in these former colonies due to its absence in the subcontinent, it kept the British people informed of the developments within the Empire helping to redefine their relationship to newly independent states.[15] With the looming independence of non-settler colonies, the British needed to maintain links and ties to these new states.

Just two years later, in 1949, the Kingdom of India declared itself a republic, ending the role of the British monarchy within the country. Even though India sought to remove the monarchy, it did not wish to leave the Commonwealth of Nations and thus requested for an exemption to be made. As part of preparing for the end of British rule in Africa, Asia, and the Americas, Britain knew that most of these soon-to-be sovereign states desired to go on their own path, becoming republics, but still desired to be part of institutional structures that they were part of previously. Due to India's request, the Commonwealth of Nations no longer required a member state to be subject to the crown.[16] Thus a new category of relations emerged. The term "Commonwealth Realms" referred to the union of states that observed the British monarch as their sovereign. Creating such institutional relationships allowed the United Kingdom to maintain greater ties with their former colonies than other colonial powers.

THE BBC

The British Broadcasting Corporation (BBC) held a legal monopoly in the United Kingdom over the airways, both television and radio.[17] The BBC had offered television since the early 1930s, but transmission was paused during World War II resulting in radio asserting its dominance until 1953.[18] The early television service primarily offered light entertainment. Yet few were able to access the service due to the cost of television sets, which by 1953 was the cost of eight weeks' wages for the average British male.[19] Television's spread was further hampered by the fact that the radio still offered quality programming on par if not better than a normal television show.

In 1932, the BBC Empire Service was created in large part due to its position as the sole station in the United Kingdom and the fact that the United Kingdom was the center of the Empire. The BBC Empire Service was intended to serve those from Britain who lived in the colonies or were English speakers. By the Second World War, the service was funded by the Foreign Office and began to use the service, now renamed the BBC Overseas Service, for propaganda.[20] By war's end, the BBC Overseas Service had become known for being Britain's voice outside of the Empire.

Television, reintroduced in 1947, had begun slowly building up an established viewer base, but the Coronation of Queen Elizabeth II in 1953 helped to double the number of TV licenses in the United Kingdom. The service, which until 1949 was only available to a 40-mile radius around London, had expanded with a new second transmitter in Birmingham, 120 miles from London, and became over the next decade accessible to every British household permanently.[21] The Service was still perceived as a lower form of entertainment, and it broadcast for only a few hours a day. Each day's television schedule was published in the *Radio Times* and, by the 1950s, there were still programs listed as audio-only — or were, as Mark Aldridge said, "radio with pictures."[22]

ROAD TO THE CORONATION

On February 6, 1952, King George VI died, and his role as Head of the Commonwealth and King passed to his eldest daughter, the new Queen Elizabeth II. In the United Kingdom, a new monarch must be presented to and accepted by "the people" in a series of events that includes oath-taking, religious anointment, and public acclamation, for example.[23] The last coronation performed was for King George VI in 1937: his brother Edward abdicated his position so quickly, George repurposed the planned coronation into one for himself.[24] This was the first coronation to be broadcast on the radio as the technology had not yet been viable for voice broadcasts at the time of the previous Coronation in 1911.

In 1937 the Coronation of King George VI and Queen Elizabeth was the first major state event covered by the BBC on television. The service covered a number of coronation festivities, although it could not televise the service itself from Westminster Abbey. As mentioned prior, the service was still in its infancy and was deemed to be illegitimate and not deserving of live broadcasting from inside the sacred hall Westminster Abbey. There was also concern about broadcasting a religious service from the Abbey alongside the non-religious activities before and after, though film cameras were allowed.[25] Even with the minimal number of potential viewers and the paradoxical viewing material, the broadcast was a success and was remembered for the next coronation.

Two committees planned such a momentous occasion. The "Coronation Executive Committee" was led by the traditional planner of coronations, the Earl Marshall. The other "Coronation Committee" was headed by the Queen's Consort, Prince Phillip, the Duke of Edinburgh. These committees were both part of the Privy Council and included senior Cabinet ministers, the Earl Marshal, the Archbishop of Canterbury, and senior members of the Royal Family. The Executive Committee also included public servants tasked with executing the Coronation. In this structure, the Executive Committee was tasked with fulfilling the decisions made by the Coronation Committee. In addition, a "Coronation Commission" also headed by the Duke of Edinburgh sought to "consider those aspects of the arrangements for the Coronation which were of common concern to the United Kingdom and other Member States of the Commonwealth."[26]

In June 1952, the BBC approached the Coronation Executive Committee arguing in favor of televising coronation events including the actual service.[27] Part of their argument was technical: the delegation asserted that television cameras required less light than the previously approved film cameras for news reels.[28] Even with the acceptance of the legitimacy of the BBC's argument, a decision was initially made to not televise the event.[29] Opponents of the BBC's position included the 78-year-old Winston Churchill, who served as Prime Minister during the Queen's ascension and Coronation, the Earl Marshall, and the Archbishop of Canterbury, who performed the ceremony.[30] The Prime Minister was a stalwart of British politics, having been first elected to parliament during the reign of Edward

VII, the Queen's grandfather, and an ardent monarchist who desired to uphold the monarchy even in the changing world.

On October 20, the Executive Committee announced their decision to not televise, but instead film the Coronation. This decision experienced quick public backlash and outcry, with some members of the Coronation Committee exclaiming that they had not been consulted. By the end of that week, Prime Minister Churchill announced a review of the ban during a speech in the House of Commons. The ban was formally lifted on December 8, giving the BBC five months to plan the television proceedings.[31] The revised plans for broadcasting the events were met with considerable good press for the Coronation, with some publications attributing the decision to the personal intervention of Queen Elizabeth II.[32]At the same time, there were concessions to the opposition. The BBC agreed not to show the Act of Consecration nor the taking of communion, as these acts were deemed too sacred to be either filmed or televised. Neither had been filmed for King George VI's Coronation sixteen years earlier. The cameras were also barred from shooting close-ups of any of the figures involved in the ceremony, with all cameras being in fixed positions.[33] American broadcasters sought greater access to the ceremony: at least one broadcaster offered to cover the cost of repairs to Westminster Abbey if they were allowed to knock a hole in the wall of Abbey to fit more cameras. The Americans were not given what they wanted within the Abbey but were granted further camera access along the procession route and facing the palace, giving the American stations a unique broadcast for those portions of the day's events.[34]

Once the televising of the Coronation was announced, a spike in sales of television sets in the United Kingdom occurred. The number of people holding television licenses, a tax that paid for the BBC television service, increased from 1.9 million in December of 1952 to 2.7 million by Coronation Day.[35] The fever for the televised Coronation rose not just in the United Kingdom but throughout the Commonwealth Realms and other British allies. The BBC planned an elaborate television relay and technical conversion process to broadcast the British signal to Belgium, France, the Netherlands, and West Germany. The conversion process was necessary due to the differing technical transmission and reception standards on the continent, and it was also the first time that a television broadcast had been shared among five countries.[36] The United States and Canada, being too far away to access BBC signals, planned to fly military jets with film reels as cargo.[37]

The plan for such an event included the problem of placement of news organizations. News organizations from Buenos Aires to Tokyo descended upon London for the historic event with 930 international reporters representing over fifty countries being given press passes by the Ministry of Works.[38] The massive influx of international reporters was not the main attraction; that was the BBC Television Service. The BBC was prepared with eight commentators, twenty cameramen, and seventy-two other people working on the program.[39] The setup was to offer full coverage of the Coronation both on radio and on television. They would cover the day's four major events: the Queen's Procession to the Abbey, the Coronation Service, the State Procession, and the Queen's appearance on the balcony at Buckingham Palace.[40] The BBC had experience with everything except the Coronation Service, resulting in a sense of nervousness for the BBC crew as they expected 300 million people to watch the Coronation from BBC sources.[41]

On the day of the Coronation, those in London received a traditional British welcome: rain.[42] Rain combined with the early television broadcast quality led to a more washed-out image, resulting in not the greatest example of television for the expanding viewer base. But the British public did not care about the viewing quality, as the sacred ancient ceremony was brought into their homes for the first time. This was the first time that British television had more viewers than BBC radio had listeners. 20.4 million adult Britons watched the Coronation through television sets, comprising of over half the adult population of the country.[43]

During the preceding weeks before the Coronation, Britons were treated to an eclectic display of the Commonwealth Nations brought to them through their television sets. Viewers were greeted by military bands and the voices of children from throughout the Commonwealth engaging in the festivities.[44] Immediately after the Coronation, the State Procession occurred with military regiments from the United Kingdom, the colonies, and Commonwealth Nations marching down the mall towards Buckingham Palace.[45] On the night of the Coronation, the BBC had a spectacle of international flair. After the Queen's speech to her realms, viewers watched a program entitled *Coronation Day Across the World*, which brought the sounds of jubilation in the Commonwealth and Europe to the United Kingdom.[46] But beyond the purpose of making Britons feel as if they were part of a global monarchy and Commonwealth, the people within that Commonwealth also experienced an ancient event brought across the seas through television.

THE MAPLE CROWN

While the Coronation of Queen Elizabeth II had many similarities to the three previous coronations of the twentieth century, the inclusion of television cameras made it a groundbreaking event. The BBC's decision to broadcast the Coronation live enabled people outside of Britain to experience the event in real time, creating a sense of shared experience across the Commonwealth and beyond. This was especially significant for the Dominion of Canada, which had a long-standing connection to the United Kingdom as one of its oldest colonies. Despite gaining independence through the Confederation, Canada remained aligned with Great Britain and kept the monarchy.

As a result of this special relationship, Canada held a significant place in the minds of the Diplomatic Corps during the Coronation. The live broadcast of the event enabled Canadians for the first time to witness the coronation of their monarch as if they were present in Westminster Abbey. During the planning process for the Coronation as part of the desire to further include the Commonwealth Realms, a decision was made to place a commentator within Westminster Abbey to broadcast in French.[47] This would have been in addition to the commentator broadcasting in English. Canadian and Australian commentators also joined the British commentating crew to flesh out the Coronation Day broadcast.[48]

A majority of Canadians also lived near the border with the United States and could receive television signals from American television markets such as Buffalo, Detroit, and Seattle. These American broadcasts competed with the relatively new Canadian CBC's own broadcasts for the same viewers. Within the United States, the two big channels, NBC and CBS, competed for viewers' attention span, and the Coronation was no different. All three broadcasters sent a technical crew to the United Kingdom to prepare for the creation of a broadcast film to be aired in North America on the night of the Coronation.[49] With the United Kingdom being five hours ahead of the eastern coast of Canada and the United States, the technical crews had from the early morning dawn in America when the Coronation started, to evening time to record, edit, and fly the telecast over the Atlantic to be broadcast. North America could have seen the BBC telecast without the need to film it, but the technical limitations of 1953 meant that the only way to have done so was to use the underwater trans-Atlantic cables and block all other use of them.[50]

All three channels had acquired fighter planes for use in this endeavor across the Atlantic, with NBC the first to leave England. Soon after takeoff, the Canberra jet, which was used by NBC, was called back due to trouble with the fuel line. NBC had edited the BBC broadcast for American television with extra visuals and commentary included, but with the plane's recall, this program would not air, and the film reels disappeared after returning to England.[51] Apparently, the BBC did not want the Americans to upstage the Canadians, especially when so many Canadians were able to receive the American broadcasts, as the CBS crew believed at the time.[52] A BBC colleague of Charles Colledge, who was second in command of NBC

News during the Coronation, confirmed it years later to him at a private conversation in London.[53]

Because of the delayed plane arrival, NBC and ABC rebroadcasted the CBC telecast, leaving CBS the only American channel to air their own original broadcast.[54] By being the first to broadcast in North America, Canada cemented itself as an integral part of the Commonwealth and helped to strengthen the bond between Canada and the United Kingdom.

CONCLUSION

In conclusion, the Coronation of Queen Elizabeth II on June 2, 1953, was a historic event that marked the first time a British monarch's coronation was televised live. The decision to televise the Coronation was not without controversy, with concerns about the appropriateness of broadcasting a religious service and the potential impact on the ceremony's sanctity. However, the decision ultimately proved to be a great success, as millions of people worldwide witnessed the historic event, which marked the continuation of an ancient tradition while embracing the modern technology of television. The Coronation also marked a significant moment in the history of the Commonwealth, as countries from around the world came together to celebrate the new Queen's ascension to the throne. The Coronation of Queen Elizabeth II remains a significant cultural and historical event to this day, a testament to the enduring legacy of the British monarchy and the importance of adaptation to the modern world.

Notes

1. Peter Hennessy, *Having It so Good: Britain in the Fifties* (London: Penguin Group, 2006), 243.
2. Miles Taylor, *Empress: Queen Victoria and India* (New Haven: Yale University Press, 2018), 69.
3. Taylor, *Empress: Queen Victoria and India*, 178.
4. Prince Alfred became a member of the German Empire due to inheriting the title of the Duke of Saxe-Coburg and Gotha, inheriting the title from his paternal uncle as his father predeceased the Duke. Ed Lowther, "Right Honourable No More: The Privy Council Quitters," *BBC.com*, February 5, 2013, https://www.bbc.com/news/uk-politics-21342555.
5. Eberhard Demm, "Propaganda and Caricature in the First World War," *Journal of Contemporary History* 28, no. 1 (January 1993): 181, https://doi.org/10.1177/002200949302800109.
6. The Royal Family, "The House of Windsor," The Royal Family, January 11, 2016, https://www.royal.uk/house-windsor.
7. The Titles Deprivation Act, 1917, 7 & 8 Geo. 5 c. 47.
8. John Cannon, "The Survival of the British Monarchy: The Prothero Lecture," *Transactions of the Royal Historical Society* 36 (1986): 159.
9. Brittany Vonow, "Royal Family's Secret Plan to Flee Britain If Nazis Invaded & Hide in Canada," *The Sun*, November 23, 2019, https://www.thesun.co.uk/news/10407739/royal-family-secret-plan-flee-britain-canada/.
10. Princess Elizabeth and Princess Margaret, *The Children's Hour* BBC, London, October 13, 1940).
11. Aletha Adu, "How Old Was the Queen during World War 2 and Did She Serve in the Military?," *The Sun*, November 11, 2019, https://www.thesun.co.uk/news/10316926/how-old-queen-world-war-2-serve-military/.
12. Max Foster, Luke McGee, and Lauren Said-Moorhouse, "Prince Philip, Duke of Edinburgh, Dead at 99," *CNN*, April 9, 2021, https://www.cnn.com/2021/04/09/uk/prince-philip-dies-gbr-intl/index.html.
13. Hamish Bowles, "Hamishsphere: A Vogue History of Royal Wedding Dresses," *Vogue*, April 25, 2011, https://web.archive.org/web/20110808021451/https://www.vogue.com/vogue-daily/article/hamishsphere-a-vogue-history-of-royal-wedding-dresses.
14. Statute of Westminster, 1931, 22 Geo. 5 c. 4.
15. Shailaja Bajpai, "The World Came Home: The History of Television in India," *The Indian Express*, July 24, 2016, https://indianexpress.com/article/entertainment/television/the-world-came-home-2932048/.
16. Editors, "Our History," *The Commonwealth.com*, 1 May 2023,

https://thecommonwealth.org/history.

17. Ronald H. Coase, "The Development of the British Television Service," *Land Economics* 30, no. 3 (August 1954): 210, https://doi.org/10.2307/3144375.

18. Brian Barker, *When the Queen Was Crowned* (London: Routledge & Kegan Paul Ltd., 1976), 129.

19. Hennessy, *Having It so Good*, 244.

20. John Tusa, "BBC World Service as a Public-Sector Broadcaster," *The World Today* 48, no. 12 (December 1992): 230.

21. BBC, "Television's Crowning Moment," *History of the BBC*, BBC.com, accessed May 5, 2023, https://www.bbc.com/historyofthebbc/research/story-of-bbc-television/televisions-crowning-moment/.

22. Tom de Castella, "How the Coronation Kick-Started the Love of Television," BBC News, May 30, 2013, https://www.bbc.com/news/magazine-22688498.

23. Editors of Encyclopaedia, "Coronation," *Encyclopedia Britannica*, May 30, 2023. https://www.britannica.com/topic/coronation-ceremony..

24. Edward Owens, "'This Time I Was THERE Taking Part': The Television Broadcast of the 1953 Coronation," in *The Family Firm: Monarchy, Mass Media and the British Public, 1932-53* (London: University of London Press, 2019), 348.

25. Owens, "This Time," 337.

26. David Torrance, "The Coronation: History and Ceremonial" (London: House of Commons Library, April 19, 2023), 24.

27. Owens, "This Time," 337.

28. Owens, "This Time," 337.

29. Owens, "This Time," 339.

30. Owens, "This Time," 338.

31. Barker, *When the Queen Was Crowned*, 132.

32. "The Coronation," *The Manchester Guardian*, December 9, 1952, http://login.ezproxy.lib.vt.edu/login?url=https://www.proquest.com/historical-newspapers/coronation/docview/479406946/se-2.

33. Jason Jacobs, "The Coronation (BBC, 1953)," in *The Television History Book*, ed. Michele Hilmes (London: British Film Institute, 2003), 71.

34. Barker, *When the Queen Was Crowned*, 134.

35. Coase, "Development," 211. Henrik Örnebring, "Writing the History of Television Audiences: The Coronation in the Mass Observation Archive," in *Re-Viewing Television History: Critical Issues in Television Historiography*, ed. Helen Wheatley (London: I.B. Tauris & Co Ltd, 2007), 347.

36. Dijk and B. Van Loon, "The Coronation of Queen Elizabeth II and the Early Days of Eurovision," *Proceedings of the IEEE* 91, no. 6 (June 2003): 977, https://doi.org/10.1109/jproc.2003.813567.

37. Richard S. O'Brien, "Television Recording 50 Years Ago: A Recollection: The Coronation of Queen Elizabeth II," *SMPTE Motion Imaging Journal* 113, no. 1

(January 2004): 28, https://doi.org/10.5594/j12328.

38. B.P. Barker, "Facilities on the Processional Route for BBC Sound and Television Services, Newsreel Reporters, BBC Observers and Press Photographers," Ministry of Works, WORK 21/303, May 1953, 3. https://discovery.nationalarchives.gov.uk/details/r/C11193664

39. Barker, "Facilities," 4.

40. Barker, "Facilities," 4.

41. Barker, "Facilities," 3.

42. de Castella, "Kick-Started the Love of Television."

43. Owens, "This Time," 347.

44. Thomas Hajkowski, *The BBC and National Identity in Britain, 1922-53* (Manchester: Manchester University Press, 2010), 101.

45. Barker, *When the Queen Was Crowned*, 193.

46. Hajkowski, *National Identity*, 102.

47. Hajkowski, *National Identity*, 102.

48. Hajkowski, *National Identity*, 102.

49. O'Brien, "Television Recording," 28.

50. Reuven Frank, "The Great Coronation War," *American Heritage* 44, no. 8 (December 1993): 2.

51. Frank, "Coronation War," 6.

52. O'Brien, "Television Recording," 34.

53. Frank, "Coronation War," 9.

54. O'Brien, "Television Recording," 34.

For War and Peace: Television in the Cold War

TYLER RIPLEY

The Cold War brought with it significant changes to the global order. Tensions that had been building since before World War Two had not been fully resolved, even with the fall of the Axis powers at the hands of the Allied forces. Instead, conflict escalated with nations realigning around the two emerging superpowers, the United States and the Soviet Union. This new alignment had high stakes: each superpower feared that the other would bring about their downfall. As a result, each side desperately tried to acquire and develop any technology or resources that could give it an advantage over the other. All the while, there were also domestic changes occurring. Military technology, developed and perfected in the secrecy of World War Two, had evolved and was starting to migrate into civilian households. Fueled by post-war consumer surpluses, new technologies like television became available for the masses, a change that, much like the introduction of other industrially produced goods, had far reaching implications, not just in domestic American politics, but also on the world stage.

In one way or another, politics have been integral in the development of just about every product ever produced. From stimulating demand to regulating its launch, politics, and its respective bureaucracy can have a significant impact on what becomes of a product. This holds true with television. In the decades preceding World War Two, television technology had been improving rapidly. Far from the primitive "shadowgraphs" that Herbert Hoover had played with as Secretary of Commerce, American television research had created mature technology that could transmit shows and even some media in full color.[1] With the technology rapidly developing, it soon came time to confront the issue of agreeing on how to broadcast it. This issue had risen to be a priority issue by 1938 when the first attempts to organize broadcast frequencies were made.[2] This early attempt, however, was met with mixed results, and no serious consensus could be reached on the issue.[3] This is, of course, the nature of the political process. The issue was soon thereafter relegated. At the time though it hardly mattered; the number of American consumers with television access was much lower than other countries like Great Britain. At the outbreak of World War Two, most of this civilian research was halted or transitioned to a more militarized version and the expansion of consumer television slowed or was even outright banned as the cathode ray tubes that made up the core of television sets were redirected towards military uses.[4]

The technology that would later make television available to the masses was also what propelled the Allied war effort.[5] Cathode ray tubes were the displays that made television possible, yet they were also needed to produce the displays for radar and sonar arrays that could detect approaching enemies from distances previously unthinkable.[6] These technological developments were critical for the war effort. The ability to be able to detect and identify incoming attacks, whether they be from submarines or airplanes, allowed Allied forces to react more quickly than before. The improvements were so dramatic that cover stories had to be created to explain the sudden newfound success of Allied pilots in Britain.[7] Officially, drinking carrot juice gave pilots improved night vision, but in reality it was the guidance from radar systems displaying on cathode ray tubes.[8] The television technology of cathode ray tubes was used extensively by both pilots and sailors during the war. German U-boats had intended to starve out the British islands, which were heavily dependent on imported foodstuffs and supplies. By preventing any merchant ships from reaching the islands, the Germans hoped that they could force Britain into submission. This, however, did not go according to plan.[9] Convoys of powerful American destroyers and destroyer escorts provided cover and defense for the slower, poorly armed and armored merchant vessels. These destroyer escorts were armed with powerful weaponry like depth charge launchers and anti-submarine mortars. New technology, like cathode ray tubes to display the location and distance of detected U-boats, allowed the destroyer escorts to hunt down and destroy them. Through the use of aircraft, equipped with surface scanning radar to detect U-boats charging their batteries at sea, the United States changed the tide of the war in favor of the Allies.[10]

The war had been hard fought and the losses from it had been devastating. Tens of millions had been killed, with industries destroyed and lives overturned. As the remaining world powers worked to draw out what post-war peace would look like, there was hope that despite the vast differences in political ideology and social order, differences could be put aside to truly form a better world. Although there was much optimism for permanent world peace, it turned out to be premature. This was not for a lack of effort as there were multiple attempts to utilize television for peaceful and cooperative purposes.

While television did not become the great unifier of the world, as many had hoped, it was not for lack of effort. One of the first attempts to make a unified television program was an international news program called *Our World*.[11] *Our World* started as part of the more optimistic period of the Cold War. Hope for a peaceful coexistence had not waned, despite political differences. As an international news show, produced and broadcast as a partnership of multiple Western and Eastern Bloc countries, the show was supposed to be one of the few unifiers that remained in an increasingly polarized world.[12] After months of preparation, a group of Eastern Bloc countries pulled out as a criticism of the Western response to the Six-Day War occurring in the Middle East.[13] Despite this setback, however, the broadcast went on as scheduled and was, at the time, the single most watched event on television.[14] Our World didn't really have the longevity that it hoped for, but it did show that international cooperation on television was not only possible, but also beneficial, at least in the short term.

As the Cold War progressed, television increasingly became a tool of political influence, but it was not without risk. Politicians could make rapid press releases and ensure almost guaranteed and instant access to the airwaves, yet the factor that made television different from radio—the image—also became a disadvantage to some. For example, many of those who listened to the presidential debate between John F. Kennedy and Richard M. Nixon on the radio believed Nixon had won, but those who watched the debate on TV strongly believed that Kennedy won. In contrast to Kennedy's youthful and photogenic appearance, Nixon had declined to wear makeup and appeared older and tired.[15] Kennedy defeated Nixon in the 1960 presidential election, partially as a result of his youthful demeanor and utilized television successfully throughout his presidency.[16] Despite being the youngest American president to be elected, JFK was not naive to the realities of the world and the Cold War, something that is clearly reflected in his Telstar speech in July of 1962.[17] Although not the first presidential speech to have been televised, with that honor going to President Truman in March 1947, nor was it the first presidential speech to be transmitted via satellite. In December 1958, President Eisenhower was one of the first to be broadcast internationally via satellite with viewers in Europe.[18]

At a time when speeches were frequently just audio or paraphrased reporting of transcripts, seeing the President of the United States appear on TV was a departure from previous trends. However, Kennedy's speech was not just on American television.[19] The power of the Telstar system allowed his speech to be transmitted to Europe where it could reach an even larger audience.[20] Speaking bluntly at first about the problems that the United States faced, President Kennedy reported on the lack of progress on the issue of Berlin and the ever-present danger of nuclear testing, but he also reminded viewers of the possibility of a diplomatic solution if the great powers could agree.[21] Clocking in at 18 minutes of screen time, the speech showed that television could open up more candid channels of communication aimed not only at one's own citizens and allies, but also opponents.[22]

Soviet leaders also began broadcasting their speeches.[23] Although a very different person from President Kennedy, Soviet Premier Nikita Khrushchev utilized television as well. In April 1964, Krushchev appeared on international television to discuss a new strategy for the Soviet Union's relations with its communist allies.[24] Despite the popular notion that the communist countries were completely aligned with one another, there were severe differences and disagreements between the nations. The Sino-Soviet split was a major disagreement between the powers, but there were also disagreements among many of the Warsaw Pact countries, many which chafed against what was seen as excessive control by Moscow.[25] Hoping to alleviate some of the tension between the nations of the Pact, and with a goal of making it stronger and better than it had been, Premier Krushchev described his plan on how to create a more equal union with the member nations. Although not as widely remembered, Krushchev's speech illustrated the geopolitical importance of television from a Soviet perspective.[26]Krushchev's ability to go on screen and describe his plans for not just how the Soviet Union would be run, but also how he would interact and negotiate with other leaders was a massive force of power projection and a significant step beyond what had previously been done.

OLYMPIC RECORDS AND MOON TELEVISION

The Olympics are an inherently political event. From the selection of the host country to the delegations who attend, the Olympics is an international spectacle in political events. Few events show this as much as the 1964 Olympic games. The Olympic games were held in 1964 in Japan and the White House as well as the American State Department wanted to create a global television broadcast of them. This would be a massive technological feat if completed successfully and would require the cooperation of multiple nations as well as multiple American government agencies.[27] The plan was to use satellites to provide live coverage of the games and transmit it to a wide range of countries around the world. Even the Soviet Union and some of the Eastern European nations of the Warsaw Pact were to participate, at least in their own discreet way. The US Department of Defense was deeply interested in the undertaking given the powerful uses that television has as well as the massive resource potential that it unlocked.Television grew fast, and soon became a force to be reckoned with in the United States. From 1950 to 1960, the percentage of homes that owned TV's grew from twelve percent to a whopping 88%, a massive increase of 76%. This increase did not go unnoticed, mainly because everyone was glued to their screens.[28] Enough high power satellites that could transmit the plethora of sports involved in the Olympics could also provide the military with a massive advantage. The entire undertaking was to be done on a nonprofit basis, despite the massive amount of resources and technological development required to even pull off the event.[29] Despite the seemingly massive requirements and challenges, the 1964 Olympic games were successfully televised live in the United States.[30] Commentators were amazed by the quality of images, despite the vast distance the signal traveled to the geosynchronous satellite stationed over the international date line, twenty thousand miles above the Pacific Ocean. Some commentators remarked that they were as good or better than footage produced in American TV studios.[31] The satellite used for this effort, named Syncom 3 and designed by the Hughes Aircraft Company, received its broadcasting signals from the Japanese Broadcasting Corporation and beamed them back to NBC for broadcast in the United States.[32] As per planned, the broadcast was also made live in other countries including Mexico, which would also be the host of the next Olympic games.[33] The success of the televised Olympics from Japan would soon be repeated.

The 1968 Olympics were hosted in Mexico and soon grew in both viewership and technology, but another historic event would supersede the Olympics, when American astronauts became the first men to land on the Moon. The Apollo 11 Moon landing brought a massive political victory for the Americans, not just in the sense of victory and achievement that comes from being first, but also from the geopolitical prowess that it proved. The success of the American method of administrative and economic structure had produced the technology that had allowed for NASA to land humans on the Moon and recover them long before anyone else. And now hundreds of millions of people around the world were going to watch history happen on television.[34]

The Apollo 11 splashdown was the epic conclusion to one of the most significant proxy fights in the entire Cold War. The space race was born out of military necessity, but brought forth a wave of technologies that would go beyond defense requirements. The space race originated at the end of World War II, when rocket scientists further developed captured German rockets. This technology had an important role in the nuclear arms race between the U.S. and Soviet militaries. Yet the search for uses of the technology, beyond just raining possible destruction upon each other, led to the exploration of new realms. The launch of the first Sputnik satellite showed that radio transmissions could be made, even from craft that were hundreds of miles up, starting a race for who could hit the next milestone.[35] Key moments in the Space Race were captured on film and broadcast. The Soviets hit the next few milestones first with the first living creature in space, first human in space, and first spacewalk in 1965.[36] Americans soon caught up with more advanced achievements like the first communication satellite and the first satellite capable of transmitting television signals, which President Kennedy used to significant effect.[37] These victories were important for national morale and prowess, at a time when individual achievements became national celebrations. All of these achievements led up to the penultimate one: landing on the Moon. A crewed Moon landing was a massive technological feat, only possible due to the huge amount of resources devoted to it. While preceding space missions had accomplished most of the tasks that would be done on the Moon mission, none of them had done them all at the same time. And almost none were as televised as the Moon mission. So significant was the Moon landing that it incidentally made some new records in the UK, with both the BBC and ITV running all night coverage of the event.[38]

On the other side of the world, in the Soviet Bloc, the responses were a bit different. For some commentators, the Soviet Union was mired in secrecy and propaganda, and truth had a bit of a different meaning.[39] Until the conclusion of the Cold War and the fall of the Soviet Union, the official story was that the space race was not real, and that it was a construct created by the Americans. Unofficially, however, the Soviet Union did have their own space program and was actively trying to develop crewed Moon missions of their own.[40] They were not just trying to make it to the Moon for its own sake but to beat the Americans in a show of Soviet political and technological power and capability.[41] This was confirmed by the Soviet cosmonaut Vladimir Komarov, who said "The U.S. has a timetable of '1969 plus X,' but our timetable is '1969 plus X minus one'!"[42] The Soviets, however, canceled their project after several failures of their Moon rockets to even make it out of Earth's orbit.[43] In the eyes of the Soviets, there is no shame in losing if you weren't actually playing.

The success of Apollo 11 was a win for the Americans as it proved the technological prestige and industrial capability of the United States, but the less successful Apollo 13 mission did not go as planned, diminishing the interest of many in the space race.[44] The first two missions had been obvious successes, as had the dress rehearsal missions like Apollo 10. The success of Apollo 13 was all but expected, yet things did not go according to plan. The launch of Apollo 13 went relatively well despite the general disinterest in the mission by the general public. Several television stations had declined to even show the launch as interest in the program had just declined by such a degree that it was considered to be insignificant and not worthy of being shown.[45] Although there was the general apathy to the launch, television did come into the picture when a televised section was produced with astronaut Lowell using a camera to provide a guided tour of the capsule for the audience of a tv show.[46] From here, the mission would continue with the expected engine to boost the capsule and its supplies into the trajectory to arrive at the Moon. Not long after Lowell's broadcast from space, a crisis occurred.[47] What would later be determined to be the ignition and explosion of a heating element within the oxygen tank had caused the oxygen tanks to rupture.[48] Not only did this result in the loss of the main source of oxygen for the crew, it also meant a loss of the majority of the electrical power of the ship.[49] The Apollo crew capsules relied on Hydrogen fuel cells which combined oxygen and hydrogen to create water and electricity, but without the oxygen, they were mostly useless. NASA and the crew of Apollo 13 had to work quickly in order to find and implement a solution. While the television cameras had been off or pointing elsewhere for the initial launch, they all quickly focused on NASA and the now struggling spacecraft.[50] The crisis soon became an international spectacle as viewers from all over the world tuned in to learn the latest updates on the damaged spaceship.[51] One might believe that such an incident would be purely negative for the Americans. They faced the real possibility of losing three highly trained astronauts, but the drama captivated and reignited interests in the Apollo program.[52] The incident also inspired a rare display of international cooperation with the Soviets and other countries offering resources and support to aid in bringing the Astronauts home safely.[53]

For all of the geopolitical crises and conflicts that occurred during the Cold War, American leaders still looked for opportunities by which they could cooperate with their ideological and military opponents. An invitation was extended to the Russians to cooperate to create a high altitude television system combining the best of both superpowers.[54] American satellites with their better technology were to be lifted into higher orbits by stronger Soviet rockets.[55] The potential for cooperation of the two largest superpowers for a shared purpose was one that had some potential as it could combine the best of both superpowers to work together for a common goal. Although cooperative rocket building, much less for the purpose of television satellites, never came to fruition, it was hoped that the political divide of the era could be crossed for a higher cause. Not until the Detente of the 1970s, however, did the tensions did relax to the point that such a collaborative mission was possible. The goal of the Apollo-Soyuz mission was simple: have an Apollo capsule and a Soyuz capsule dock in space and let the astronauts fly around together to show their newfound political cooperation. The entire mission took several years of planning and preparation but yielded a powerful political result. This, of course, could not have been realized had the event not been filmed, which it was, and subsequently broadcast around the world in real time.[56] Astronauts and cosmonauts were pictured exchanging gifts of food and commemorative items in space as they floated around the joint spaceships.[57] They remained in this joint state for a few days, enjoying the blissfulness of space, free from concern about domestic political issues or worldly problems. Space travel in this era, however, did not have the unlimited endurance of more modern space stations. The astronauts and cosmonauts, having finished their breaking of the tensions in their high altitude handshake, had to return back down to their respective nations to face the worldly political issues of the day.

Television, despite being easy to overlook, took on an important part in the politics of the Cold War. The ability to use it to send political messages, display technological and national successes and show international cooperation for a higher cause all meant that its role in the Cold War era was irreplaceable. From pivotal moments like Kennedy's telestar speech that could be seen all the way over in London to the first televised Olympic games, showing the most skilled and strongest of every nation, to the space race and the fiercely competitive environment that it brought, television was there for it all.[58] Even as the Cold War was left behind, the many uses that television carved for itself continued to be used. Olympic games continued to be broadcast as were space launches, while politicians continued to give speeches that were televised with some being made purely as TV addresses. The Cold War and its proxy battles left their mark on the world as the petty challenges, wars and near constant threat of nuclear annihilation made every decision that much more valuable. And television was there to capture it all. Without the Cold War, television probably would have evolved a little differently, but without television, the Cold War would have been just another page in a history book.

Notes

1. Coleman Rosenberger, "Television in the Brave New World," *The Washington Post*, Feb 15, 1942.
2. Rosenberger, "Television in the Brave New World."
3. Rosenberger, "Television in the Brave New World."
4. Claudia Reinhardt and Bill Ganzel, "TV Turns On" *Living History Farm*, (n.d.) https://livinghistoryfarm.org/farming-in-the-1940s/tv-turns-on/.
5. Reinhardt and Ganzel, "TV Turns On."
6. Reinhardt and Ganzel, "TV Turns On."
7. K. Annabelle Smith, "A WWII Propaganda Campaign Popularized the Myth That Carrots Help You See in the Dark," *Smithsonian.com*, August 13, 2013.
8. Smith, "...Carrots Help You See in the Dark."
9. Christopher Klein, "How German U-Boats Were Used in WWI—And Perfected in WWII," *History.com*, March 21, 2022.
10. A. Timothy Warnock, "Air Power vs. U-Boat," US Department of Defense, https://media.defense.gov/2010/May/25/2001330266/-1/-1/0/AFD-100525-066.pdf, 1999.
11. "The First Global Live Satellite Television Production - 1967," *History.info*, accessed May 3, 2023.
12. "The First Global Live Satellite Television Production - 1967."
13. "The First Global Live Satellite Television Production - 1967."
14. "The First Global Live Satellite Television Production - 1967."
15. Editors, "Kennedy and Nixon square off in first televised presidential debate," *History.com*, November 24, 2009.
16. Editors, "Kennedy and Nixon square off."
17. Murrey Marder, "Kennedy Hope of Pact on Berlin and Tests is Relayed by Telstar: Europe to America, Via Telstar," *The Washington Post, Times Herald*, Jul 24, 1962.
18. Laurie Austin, "President Truman's Media Milestone: The First Televised Speech from the White House," *The Unwritten Record*, National Archives and Records Administration, October 4, 2022, https://unwritten-record.blogs.archives.gov/2022/10/04/president-trumans-media-milestone-the-first-televised-speech-from-the-white-house/. Jessica Macneil, "1st radio broadcast transmitted from space, December 19, 1958," EDN, December 19, 2019.
19. Marder, Staff Reporter, "Europe to America, Via Telstar."
20. Marder, Staff Reporter, "Europe to America, Via Telstar."
21. Marder, Staff Reporter, "Europe to America, Via Telstar."
22. Marder, Staff Reporter, "Europe to America, Via Telstar."

23. "Challenge of Change," *The Washington Post, Times Herald* (1959-1973), Apr 14, 1964.
24. "Challenge of Change," *The Washington Post, Times Herald* (1959-1973), Apr 14, 1964.
25. "Challenge of Change."
26. "Challenge of Change."
27. "U.S. Pushes for Full TV of Olympics," *The Washington Post, Times Herald,* July 5, 1964.
28. "88% of U.S. Homes have Television Sets," *The Washington Post, Times Herald,* Aug. 31, 1960.
29. "U.S. Pushes for Full TV of Olympics," *The Washington Post, Times Herald,* July 5, 1964.
30. Jack Gould, "Sharp TV Pictures of Olympics Relayed Live to U.S. by Satellite," *The New York Times,* October 10 1964.
31. Gould, "Sharp TV Pictures."
32. Gould, "Sharp TV Pictures."
33. Gould, "Sharp TV Pictures."
34. Robert J. Samuelson, "Satellite TV Record Shattered: Nine Major Programs," *The Washington Post, Times Herald,* July 25, 1969.
35. Becky Little, "The Soviet Response to the Moon Landing? Denial There Was a Moon Race at All," *History.com,* July 11, 2019.
36. Little, "Soviet Response."
37. Marder, "Europe to America, Via Telstar."
38. Jenny Rowan, "How the UK Watched the Apollo 11 Moon Landing," National Science and Media Museum blog, July 18, 2022. https://blog.scienceandmediamuseum.org.uk/how-the-uk-watched-the-apollo-11-moon-landing/
39. Little, "Soviet Response."
40. Little, "Soviet Response."
41. Little, "Soviet Response."
42. Little, "Soviet Response."
43. Little, "Soviet Response."
44. Gary Arndt, "Apollo 13," *Everything Everywhere,* December 14, 2022.
45. Arndt, "Apollo 13."
46. Arndt, "Apollo 13."
47. Arndt, "Apollo 13."
48. Arndt, "Apollo 13."
49. Arndt, "Apollo 13."
50. Arndt, "Apollo 13."

51. Arndt, "Apollo 13."
52. Arndt, "Apollo 13."
53. Arndt, "Apollo 13."
54. "Russia Invited to Join World Television Setup: Several Dozen Needed," *The Washington Post, Times Herald*, Sep 25, 1962.
55. "Russia Invited."
56. "The Mission of Apollo-Soyuz," *C-SPAN*, accessed May 3, 2023.
57. "Apollo-Soyuz."
58. Marder, "Europe to America, Via Telstar."

Lascia o Coppia

The Legacy of Fascism for Postwar Italian Television

ANTHONY SALISBURY

The Second World War—perhaps the most significant conflict between the world's three major 'modern' ideologies- liberalism, socialism, and fascism—led to an era of unprecedented reconstruction.[1] Defeated fascist nations had to confront questions of the conduct and culpability in the history and memory of the war. In Italy, this was complicated by the nature of the revolt that removed the fascist regime from power. Italy surrendered to the Allies on September 8, 1943, but fascist leader Benito Mussolini fled to German-occupied Northern Italy to lead a puppet state there. After his death in April of 1945, the monarchy, which had nominally held power throughout his reign, returned to political power, despite the distaste of the republicans and socialists, both of whom viewed the monarchy as responsible for the rise of fascism. Yet, despite the shared ground of opposing fascism and monarchism, the republicans and socialists found themselves at odds concerning the nature of the government that would follow the autocratic regimes. They vied to determine how to remake the Italian state and domestic policy in the face of this multi-layered conflict. Italians across the political spectrum were concerned about the seemingly inevitable continuation of conflict between the two dominant ideologies, liberalism and socialism. The economies of most of Europe had been devastated by the war, between the damage of fierce fighting across much of the continent, and the severing of trade ties between nations for the better part of a decade. The era of reconstruction allowed many of the combatants to address the legacies of the war. Italy was spared the fate of occupation or partition, but new problems arose, such as how to grapple with the legacy of fascist ideas and legislation. The solutions were not clear: the authoritarian controls wielded by Mussolini's regime could be greatly useful for a new government, for example, but at what moral cost? While the propagandistic efforts of the regime may have failed to maintain control of the Italian people, the system that enabled such efforts remained intact. For example, the newly flourishing field of television already had an established state-run monopoly, as did radio. This paper evaluates the legacy of television as a source of political propaganda as pioneered during the fascist era to show how post-war politicians used fascist legislation as a tool for state-building in a fragile era, entrenching television's influence over the postwar political climate and the historical legacy of fascism in Italy.

The expansion of television as a medium of communication began long before the Cold War, and much like radio, the invention of television opened up a new world of information control and influence. Indeed, television proved a more controllable medium, thanks to the different production conditions that required more intensive technological preparations behind television productions, and the greater need to pre-plan what goes in front of a camera. Additionally, television production was a fairly expensive process, making it harder for amateur productions to exist, let alone get on the airwaves to the people. Thanks to these and other factors television served as an easy medium for the state to dominate for their own purposes, especially once the fascist legislation established a state-run monopoly.

Given the rise of authoritarian ideologies of the time, specifically fascism in nations like Germany and Italy, this opportunity for a new, powerful medium of propaganda was harnessed. The first televised program in Italy, for example, was an experimental broadcast of a speech given by dictator Benito Mussolini, sent out with the hopes of inspiring nationalism in the people of Italy, and making Mussolini seem like a strong leader. The speech was broadcast by URI (*Unione Radiofonica Italiana*), a fascist-era corporation that would shift in form over the next two decades to become RAI (*Radio Audizioni Italiane*) in the republican period, albeit without significant changes to the foundational fascist legislation that had created it.[2]

Patterns of consumption in Italy also shaped television's usefulness in propagandizing. Radio was wide-spread, reaching Italians predominantly used in their homes at the time. Meanwhile, television's relative scarcity meant that consumption of television was often communal. People often gathered in bars around TV sets to watch new programming, a trend that would last well into the post-war era.[3] This, additionally, made content like speeches perfect for television in the early years—thanks to communal watching, it became much easier to reach a large audience on television, especially for messages designed to give the people a unifying drive in the name of fascism.

This was just one example of a general pattern in fascist systems of control-influence in the arts. Mussolini, much like Hitler, attempted to channel art in Italy to serve the state.[4] Mussolini found allies among the proponents of Italian Futurism, an art movement which emerged around the same time that the fascist party had been founded. Futurism focused on notions of progress and strength, stating that the future was well within reach for those willing to make certain sacrifices. As many notable Futurist artists held sympathies towards Mussolini's ideals, it was easy for the fascist state to portray itself as the beacon of modernity and progress, using the new avant-garde art style as an unofficial aesthetic. Reminiscent of Goebbels' efforts in Nazi Germany, they delivered messages of duty, loyalty, and masculinity, along with the story of 'mutilated victory'- a victory won by Italians in the First World War, but stolen away by people like the British and French. While the existing culture and structure concerning fine arts in Italy made it difficult for fascists to directly exert control, they attempted to appropriate exhibitions to push fascist narratives and themes.[5]

The fascist regime in Italy used displays and exhibitions beyond art, including achievements in a number of fields, from aeronautics to public health, to demonstrate the government's strength. These campaigns served to build up the image of the fascist government being well-connected and well-prepared in connection with the Futurist narrative that fascism was the ideology of progress. Through this, the Italian government took on the role of patron for any number of industries and their showcases, establishing a formula of exhibitions as propaganda that included massive talent, and often merged entertainment with propagandizing. Even the exhibitions themselves took on the trappings of Italian Futurism, with an avant-garde visual approach and a blurring of the divide between the audience and the show, attempting to form a sense of identification on the part of the audience member with the future the fascist party sought to create.[6]

Fascist cultural propaganda largely failed to reach Italians and change their world views. Italians relied more on their personal experiences with the world around them to inform them, rather than relying on the government message about progress and politics. Fascist popular culture focused on creating national unity under the regime but failed to do so predominantly due to inherent indifference among the Italian populace. Fascist propaganda succeeded in making Mussolini a larger than life figure, a mostly insignificant result when it came to controlling the populace.[7] Additionally, American literature and cinema was seen widely as a beacon of modernism and imagination. While the fascist politicians strove to invigorate domestic arts, foreign creations captured the attention of Italian audiences throughout the fascist era.[8] As such, films and television broadcasts pushing fascist propaganda fell by the wayside, replaced by the more adventurous, and often perceived as less politically loaded, cinema from the United States.[9]

After the collapse of the fascist regime in 1943, the newly formed republican government faced the challenge of reconstructing the state and nation. Their task was complicated by the ideological diversity of postwar political factions. The partisan revolt that had destroyed the regime was an umbrella movement encompassing anti-fascist groups, who were contending for control of the government. Chief among them were two parties; the Christian Democrats, a party inspired by the ideals of western liberalism, and the PCI (*Partita Comunista Italiana)*, a Communist party sympathetic to Stalinism in the Soviet Union. In addition, Italy was a deeply faithful nation of Catholics, and the Vatican exercised significant influence over Italian politics, often favoring the Christian Democrats, due to their more 'traditional' values, such as priority on family and faith.

Additionally, thanks to their heroism during the Second World War, anti-fascist partisans were hailed as heroes by a wounded nation. Across the political spectrum, people rose up to oppose Mussolini's regime. Resistance groups became heroes like the romantic-era rebels often hailed as the 'Founding Fathers of Italy', such as Garibaldi or Mazzini from the revolutions of 1848, both of whom had anti-fascist groups named in their honor (the Garibaldi Legion and the Mazzini Society, respectively.) The resistance movement and the destruction of the regime were celebrated as the birth of a new age for Italy. The resistance even had a national holiday dedicated to them—Liberation Day, celebrated every April 25th, on the anniversary of the declaration of general revolt against the Italian Social Republic, a German-controlled state in Northern Italy.

The ideological struggle of the Cold War raised the stakes of domestic political conflict between the Christian Democrats and the Communists. As the Cold War intensified, superpower conflict pulled small European states into the ideological orbits of Cold War liberalism and communism. States like Italy, which had strong domestic support for communism, struggled to define their own paths against foreign pressures. In Italy, the Christian Democrats were the most powerful group due, in part, to support from the Vatican, as well as the pillars of the western liberal alliance, the United States and Britain. By contrast, the communist party appealed to those who had not historically held power, due to their heroism in the war and their left-wing populist ideals that focused on improving the economic situation of the working class. Communist leader Togliatti found the conditions ripe for a communist revolution in Italy. But he relied on the Soviet Union for advice and support and, between 1945 and 1947, Stalin was more interested in conciliation with the western alliance.

Radio had been widespread and popular in Italy, and fascist-era legislation established a state-run monopoly on broadcasting under direct control of executives answering to parliament. Thanks to this precedent, the Christian Democrats realized that radio and television were important methods to reach the people of Italy, and began the efforts to convert both mediums into a source to undermine communist support.[10] The Christian Democrat party effectively established control over the former broadcasting system, with the support of the intellectual elite involved in media production and loyal party members that held most high positions of management.[11] This unilateral control by the Christian Democrats enabled them to edge the communists out of broadcasting and give themselves all the power over decisions of content and practices going on the air. As such, the new liberal republican administration rebuilt the broadcasting system not on the liberal principles of freedom of information and expression of a diversity of viewpoints, but rather on the authoritarian model of the former fascist regime. The legislation passed in the early years of the postwar era to govern Italian broadcasting is nearly identical to the fascist legislation that preceded it.[12] With this perpetuation of legislation from the fascist era, the Christian Democrats established television's purpose: to disseminate political propaganda that would recreate Italy as a liberal Catholic nation in the postwar period. While the control of the airwaves was levied mostly to combat any potential resurgence of fascist ideas, it also solidified Italy's political position in further ideological conflict — an ally of the west, despite not insignificant support for the Soviet Union.

The Catholic Church recognized the advantages of television as a method of propagandizing as well. Pope Pius XII declared that television would offer a great chance to "spread the message" across the airwaves or, in other words, that television as propaganda would benefit the political, cultural, and social interests of the Catholic Church. As such, they were more than happy to support the Christian Democrats' efforts to rebuild social structures and the broadcasting system in particular around Catholic values of family and Christian morality. In alignment with the Church, the Christian Democrats instituted certain guidelines for broadcasts, considered standards of 'self-discipline.' Any reference to sexuality was off-limits, and comments which could be deemed insulting or derogatory of the family norms of Catholicism were heavily censored. These rules only changed when Filiberto Guala, one of the upper executives of the RAI and a loyal Christian Democrat, retired, allowing newer executives to pursue a new, more liberal direction for the network. Guala went on to become a Catholic monk, which highlights the extremely close ties between church and party.[13] Thus, despite the collective hopes of people across Italy for a new age of free media in the public sphere, the actual politics of the RAI's television service told a very different story. With the new broadcasting system in place, the Italian state-run television network, *Radiotelevisione Italiana* (hereafter referred to as the RAI, a nickname derived from the company's previous name, *Radio Audizioni Italiani*) began transmissions on January 3, 1954, a date billed as 'the Day Television began.'[14]

In addition to television, other media were subject to the restrictive policies of the Christian Democrats and the Catholic church. The Italian government ensured that the Catholic Church took on a significant role in Italian censorship of the film industry, both in funding the Catholic Cinema Centre and in ensuring that members of the CCC were allowed access to the meetings of film censors. Representatives of the CCC exerted considerable pressure on the censorship process. In one notable instance, Giulio Andreotti, an influential bureaucrat involved in many censorship decisions, admitted to the secretary of the Vatican City State that he chose censors who he knew aligned as closely as possible to Catholic moral expectations.[15] Thanks to these ties and the previously mentioned code of "self-discipline," the Vatican was an ever-present force in the politics around television and film creation. Even when they were not directly involved in the creation or censorship of content, their control could be felt more personally for the Italian population. According to Trevari Gennari and Dibeltulo's work with the Italian Audiences Project, many people in Italy felt that they were being forced to make a choice by Catholic figures in their communities: to avoid content of which the Church did not approve or be shunned by the Church.[16] In a deeply religious country like Italy, being shunned by the Catholic church was unthinkable for many people. In many ways, this created an informal form of censorship that only served to further Christian Democrat control over media during the era.

The RAI set out to be the only source of information the Italian people would need, by providing news broadcasts from around the country, sports programs, documentaries, and cultural showcases predominantly produced and filmed in Italy. While it offered foreign content, such as news from the UK and France, it primarily served as a means to reinvigorate local industry, especially given the effort to promote Christian Democrat ideals. As such, unlike during the fascist era, the RAI did not rely as heavily on American media to grab the people's attention. In fact, the opposite was true: when the United States Information Service (USIS) attempted to broadcast *Voice of America*, (a propagandistic radio network espousing the American ideals of liberation, freedom of thought, and "peace," while condemning Soviet imperialism) VOA broadcasters immediately noticed that RAI's listening figures were far higher than those of *Voice of America*. As such, the USIS labeled RAI as a critical cultural influence to be used to further the American message, and they approached RAI offering to produce content for RAI's exclusive use.[17] Some content included American-made films and television programs, while other pieces were comprised of news items created specifically for RAI's anchors to promote on the air without acknowledging their origin. The RAI did not take much interest in the deal but eventually agreed after realizing their shared goal in weakening the Communist political presence in Italy. The arrangement ultimately benefited both parties, because American propaganda was conducive enough to pro-Christian Democrat messages, and the deal both provided free content to fill a demanding schedule and a means for the Italian government to create closer ties to their Western allies.

Beyond the direct interventions of *Voice of America*, American influences were still incredibly prevalent in Italian broadcasting. In television, many popular programs were derivative of American counterparts, such as *Lascia o Raddoppia*, (*Leave it or Double it*) a program inspired by game shows such as *The $64,000 Question.* However, even these programs served the propaganda machine of the RAI — the trivia questions in many of the shows were worded as to further the ideology of the Christian Democrats. As plenty of communists across Italy were beginning to notice, game show content disseminated a particularly liberal view of the world; contestants were hardly getting quizzed on Marx's *Capital* or Gramsci's political theory, as much as they were on the history of soccer, or American literature.[18] Additionally, one of the most famous television hosts of the era was an Italian-American man known as Mike Bongiorno. Bongiorno had served as an anti-fascist partisan in World War II and had a series of peculiarities that made him a cultural phenomenon. He was considered a role model for the white-collar workers of Italy and represented the aspirational aspects of television: he boasted a well-dressed appearance and seemingly classless air that contrasted with the presenters of the fascist era, and his shows claimed to promote a free exchange of ideals and discourse.Bongiorno's popularity, method of communicating, and television appearances made him a cultural touchstone so well-known that the famous Italian writer and philosopher Umberto Eco put out an essay entitled *Phenomenology of Mike Bongiorno.*

The people of Italy were well aware of the political biases in the media, and by the late 1950s, the PCI called upon the RAI to improve the diversity of opinion, and freedom of speech on television. Calling for a 'parliamentarization' of the RAI, the PCI demanded more of a presence both in the corporation (in management) and on the air. In response to this, and inspired by the televised presidential debates between Nixon and Kennedy in 1960, the RAI launched a new program hailed by critics as the first steps to democratizing television, ushering in the era of tele-politics (the use of television for political discourse) in republican Italy.[19] In actuality, *Tribuna Politica,* (*Political Gallery*) allowed the Christian Democrat government and government-aligned managers of the RAI, to deflect some of their most common political criticisms pertaining to the biases of the programming, while still maintaining, and in some ways furthering their use of the format as a cultural and social method of control. By expanding the offerings of political programs, the Christian Democrats gave themselves *carte blanche* to platform themselves on the airwaves first and foremost, while making pretense to democratizing television for parties like the communists by offering them nominally equal access. While this was a great expansion on the access of the communists to television, the Christian Democrats still controlled the station in most significant ways.

Republican-era audiences were also concerned about political influence in the media. Viewers worried about the overt politicization of television, as seen in letters to outlets on both sides of the political dichotomy. Many letters, especially in communist party newsletters or periodicals, decried the clear bias toward the Christian Democrats, but, as important, audiences on both sides were predominantly concerned that many programs were boring and otherwise unappealing.[20] Content seemed too clean and flavorless for the audience's standards, and even Bongiorno's rampant popularity was hardly enough to liven up the entire schedule. In part, the audience's boredom with RAI's broadcasting was due to an internal crackdown on content that was critical of established social norms and the government, in an effort to ensure that postwar reconstruction proceeded smoothly. For example, satirical programming, such as the farces of Dario Fo, who, along with his wife Franca Rame, were the victims of routine censorship on the part of the RAI. Supporters of satirical programming argued that the RAI targeted Fo and Rame because they highlighted "the deceptive and falsified version of contemporary world offered by the RAI."[21] The disparity between official RAI depictions/representation of the world and Fo's subversive humor exemplifies how the government instrumentalized RAI to reconstruct the Italian nation in the image of an idyllic republic, while shutting down dissenting views or criticisms.

Christians Democrats exercised this control not just over domestic politics, but also to rewrite their reputation for the world stage. Their vision of modern liberalism depicted Italy as a friendly helper in the economic development of the world, rather than the imperialist oppressor, as many nations saw it during the fascist era. During Italy's 'economic miracle,' or the great economic boom that encapsulated Italy's post-war economic recovery in the mid-1950s, a large number of Italians emigrated to the areas that had suffered the most under fascist-era Italian imperialism, such as Greece, Ethiopia, and Libya.[22] Early Italian television reported on the activities of these migrants in documentaries following their lives abroad. These programs created a sort of international view of the Italian people, and crafted a new narrative that characterized Italian migrants not as colonizers, imperialists, or oppressors, but as working-class collaborators, modernizers, and managers. The Christian Democrats used this labor-centric view of migration to former colonies, in the form of the new Italian worker, to maintain an appeal along the same lines the communist party had used to gain support from labor communities.

With this rewriting of their reputation and place in the world, Italian television further perpetuated the aims of the Christian Democrat party, both in internal politics, and foreign affairs. Not only had they undermined a significant cause for communist support, they had further encouraged a form of immigration that had greatly benefited the Italian economy: for an economy still recovering from the war, the remittances sent home by migrants to support their families injected much needed foreign capital into the Italian economy. Additionally, the documentaries seemed to draw a clear line between the Italians watching the programs and the Italians creating this new reputation — even interviews with the migrants were re-recorded and recounted by a narrator who read all of their responses for them. The pragmatism of the action is apparent; by endorsing a new image of the heroic Italian worker in the new democratic age, they reap much benefit. But by creating a disconnect between audiences and migrants in these ways, they might also prevent the risk of a generation of workers leaving the nation for foreign locales.

All in all, the RAI proved over the first decade of its television broadcasts that it was a cold, calculating bureaucracy and an ever-loyal tool of the Christian Democrats, and. Under the party's monopoly, fascist legislation and the use of pre-republican formulas to inform, influence, and indoctrinate audiences were instrumentalized to project Catholic moral standards and Democratic ideals. While the Italian people were aware of media bias and witnessed censorship first-hand, they had little chance to truly oppose or change the structures that enabled such authoritarian politicking in the first place. Similarly, the communist party, muzzled first by Stalin then the arrival of the American Marshall plan, and constantly undermined by the Christian Democrats, were never able to wield the power to dismantle the Christian Democrats' authority over television. Over the course of the era, communist support continued to slip, not only because of the weaponization of television, but the 'economic miracle,' which improved the standards of living for most Italian workers. The role RAI played in undermining the once-popular Communist Party demonstrates how important television became in both domestic politics and the emerging Cold War. By using fascist-era legislation to win the race to seize control of the airwaves, the Christian Democrats had all but assured their political dominance in the post-war era.

Notes

1. Azar Gat, 'The Major Contenders of Modernity: Liberalism, Socialism, Fascism', *Ideological Fixation: From the Stone Age to Today's Culture Wars* (New York: Oxford University Press, 2022), chap. 4, https://doi.org/10.1093/oso/9780197646700.003.0004, accessed 6 Apr. 2023.
2. Giulia Guazzaloca, "The Revolution That Never Was: Rai-TV and the launch of *Tribuna Politica*," *Modern Italy* 14, no. 3 (August 2009): 276.
3. Mark Hayward, "Good Workers: television documentary, migration and the Italian nation, 1956–1964," *Modern Italy* 16, no. 1 (February 2011): 1
4. Mino Argentieri, *La Censura nel Cinema Italiano,* (Roma: Editori Reuniti, 1964): 39.
5. Marla Stone, *The Patron State: Culture and Politics in Fascist Italy,* (Princeton, NJ: Princeton University Press, 1996), 128-131.
6. Stone, *The Patron State*, 128-131.
7. Edward Tannenbaum, *The Fascist Experience: Italian Society and Culture, 1922-1945,* (New York: Basic Books, Inc. Publishers, 1972), 213-215.
8. Jane Dunnett, "Anti-Fascism and Literary Criticism in Postwar Italy: Revisiting the 'mito americano'" in *Culture, Censorship and the State in Twentieth-Century Italy*, eds. Guido Bonsaver and Robert S.C. Gordon, (London: Legenda, 2005), 109-113.
9. Victoria De Grazia, "Mass Culture and Sovereignty: The American Challenge to European Cinemas, 1920-1960," in *The Journal of Modern History* 61, no. 1 (1989): 56.
10. Guazzaloca, "The Revolution that Never Was," 278.
11. Garofalo, *Political Audiences,* 34-35.
12. Italian Parliament, *La Radiotelevisione in Italia; Legislazione Documenti Parlamentari dal 1958 al 1963,* (Milan: Stadium Centro studi economico sociali, 1963), 17-23.
13. Garofalo, *Political Audiences,* 35-36.
14. Servino Sernesi, "Da Oggi la televisione inizia," *Radiocorriere TV,* January 3, 1954, 3.
15. Daniela Trevari Gennari and Silvia Debeltulo, "'It Existed Indeed...it was all over the papers:' Memories of Film Censorship in 1950s Italy," *Participations* 14, no. 1 (May 2017): 238-239.
16. Trevari Gennari and Debeltulo, "'It Existed Indeed,'" 240-241.
17. Simona Tobia, "Did the RAI buy it? The roles and limits of American broadcasting in Italy in the Cold War," *Cold War History* 13, no. 2 (2013): 175-177.
18. Garofalo, *Political Audiences,* 8-9.

19. Guazzaloca, "The Revolution That Never Was," 275-276.

20. Garofalo, *Political Audiences,* 46-47.

21. Luciana d'Arcangeli, "Dario Fo, Franca Rame and the Censors" in *Culture, Censorship and the State in Twentieth-Century Italy,* eds. Guido Bonsaver and Robert S.C. Gordon, (London: Legenda, 2005), 158-161.

22. Hayward, "Good Workers," 3-4.

Nuclear Screening

A Look Into The Operation Tumbler-Snapper Television Broadcast

JEREMIAH ALBERT

On April 22, 1952, "David" woke up and prepared for the day. David had just finished moving with his family into a new farmhouse in Yucca Flat, Nevada. His wife "Susan" started her morning by feeding the several pigs, sheep, and goats they raised. Together they had six children, three boys and three girls. The girls Kelly, Marjorie, and Dot arose to do their chores, and were primarily occupied in the basement cleaning the house. The boys Nicholas, Louis, and Michael, having completed their early morning tasks, flocked into the living room where David was unboxing their new television set. With the morning coming to a close, it had the making of a beautiful day. Unfortunately, none of them knew the danger they were in, nor could they have done anything to get out of it. In an instant, with a flash of blinding light, the family, the house, and the animals outside disappeared; meanwhile many miles away crowds of people looking on through binoculars cheered and rejoiced at the complete and utter destruction of the family, house, and animals.

The disaster was the intended result of a series of tests involving the detonation of nuclear bombs conducted by the United States government at Yucca Flat, part of an operation code-named Tumbler-Snapper. This operation explored different bomb types, compositions, and delivery methods. The key aspects of the tests were to determine the reliability and effectiveness of the bomb blasts.[1] By 1952, these types of tests had become routine, and a standard operating procedure had been created to administer the tests. However, the test conducted April 22, 1952, introduced a new variable: people. The variable was not because of the family, because none of the inhabitants of the house were alive. Instead, they were mannequins dressed just like regular people. The mannequins simulated humans in an explosion without real humans being present. The same could not be said for the animals put under anesthesia outside the home, but more on that later. Instead, the government this time had invited civilian spectators to watch the detonation and the disintegration of the homestead.

The bomb that dropped April 22, 1952, code-named "Charlie," was the third bomb of the series (the first two had been "Able" and "Baker") and detonated exactly as planned. Charlie was different from the other tests: this third "Shot" in Operation Tumbler-Snapper had the distinction of being the first live broadcasted nuclear explosion.[2] Shot Charlie would be the closest real human beings had been to a nuclear explosion since Hiroshima and Nagasaki.[3] Not for seven more years would the public get the chance to watch and experience a nuclear explosion live once again.

This chapter explores the role of television in shaping public opinion of American nuclear testing in the 1950s. For many Americans, filmed footage of the wartime bombings of Hiroshima and Nagasaki had been the images that defined nuclear power. In the postwar period, nuclear weapons instilled fear among Americans. Operation Tumbler-Snapper unintentionally helped change the view of the public. This chapter draws primarily from The United States Nuclear Test Personnel Review (1982) of Operation Tumbler-Snapper. The explosive results of Charlie have been studied in depth to further develop nuclear weapons for the United States arsenal. Other nuclear scholars such as James Mahaffey, Scott Sagan, and Kenneth Waltz offer information of the origins on nuclear weapons and energy but lack insight into what Americans thought of early nuclear tests. Second, it builds upon work in television studies that has reconstructed the role of this new media in shaping American public opinion. Television studies scholars, such as Helen M. Davis, and Bernard Brodie, have shown how television can impact the public's opinion. This chapter shows that the American public opinion on nuclear weapons changed because of this live broadcast. Furthermore, this research makes clear that spectators' opinions of nuclear energy and weapons differed based on how they watched Shot Charlie explode.

NUCLEAR NEVADA

The American purpose for nuclear weapons changed from an offensive use to a defensive use after 1945. Following the end of the Second World War the United States had a monopoly on nuclear technology, until 1949 when the Soviet Union manufactured and tested their first nuclear weapon.[4] With United States-Soviet relations worsening (because of the international competition between communism and capitalism), Americans were confronted with the possibility of nuclear war. Since the U.S. no longer had sole ownership of nuclear bombs, government leaders decided to make sure America had the most developed ones.[5] Over the next years the U.S would develop and test bombs secretly in the American desert. The desert had been home to nuclear bomb tests since the creation of the nuclear bomb itself. It offered a unique atmosphere that few other locations offered. Most importantly it offered seclusion. Not many people lived in the desert and those sparse populations could be easily persuaded to move. It also helped that these barren wastes were located well within the homeland territory of the U.S. which helped conceal them from foreign observers. Unlike future test sites conducted in the ocean, the desert offered the ability to test detonations on structures that could simulate cities.

The need for nuclear tests on structures is primarily a result of cultural differences. When the two nuclear bombs were dropped on Japan, they detonated above large Japanese cities. At the time, Japanese cities, buildings, and equipment were constructed from different materials compared to other nations.[6] For these reasons, the U.S. government built "doom towns" in the desert to mimic Anglo-American and European industrialized locations and structures.[7] These towns, often "inhabited" by mannequin family units, were used to see the effects of the blasts at various heights, distances, and many other alternating conditions. The towns often contained houses that mimicked a structure/home one may find in any American neighborhood. Parked outside of these homes, including the ones at Yucca Flat, were cars, tanks, personal carriers, artillery pieces, and many other military vehicles and equipment. After the detonation of a bomb, blast recovery crews could walk across the destruction to examine the devastation and report the effects to give accurate information on what might happen to civilian populations and military equipment in the United Sates if a nuclear detonation occurred in an American city. The uncertainty of the future encouraged doom town development and nuclear testing.

ATOMIC TESTING

Operation Tumbler-Snapper was ultimately justified as a way to "diversify and strengthen the U.S Nuclear arsenal."[8] The Soviet Union's own nuclear weapon development and the ongoing war in Korea also stirred interest in advancing the Department of Defense nuclear abilities. The significant number of American troops in Korea, as well as the U.S. military brass' realization that many European nations were unwilling or unable to "develop effective military capabilities" explained the need for weapon alternatives.[9] Indeed, after the Second World War, many nations' militaries, defensive infrastructure, and general industry were weakened and would require decades to repair. The United States felt not only compelled but forced to further weapon development because of the isolated situation in which it found itself.

At this point in nuclear bomb development the focus of the biggest and baddest bomb started becoming less important. The shift away from the biggest bomb was primarily due to the large number of non-military targets destroyed in the 1945 campaign. Although larger nuclear bombs were indeed produced in the future by both the U.S. and the U.S.S.R, the United States sought versatile nuclear weapons that could do more than just level cities. Thus, Operation Tumbler-Snapper was a steppingstone into the development of strategic use nuclear weapons.[10] Strategic nuclear weapons were intended to be small but effective. For example, one type of strategic nuclear bomb could have destroyed a bunker or weapons depository in the middle of a town or city. Strategic bombing provides a way to destroy military targets with potentially fewer civilian deaths. Today, a large part of the United States nuclear arsenal is made of these strategic nuclear devices.[11]

The American public opinion of nuclear testing during changed very rapidly after the war. Initially in 1945, thirty-seven percent of Americans believed they were in "real danger" of a nuclear attack.[12] Five years later in 1950, when asked if Russia would use the atomic bomb against nuclear cities ninety-one percent of the population agreed.[13] Most American citizens believed that the United States was at increasing risk of a nuclear attack. With this possibility in mind, many citizens tacitly supported nuclear tests and development. This is the context in which the 1952 nuclear tests took place.

EARLY OPERATIONS

At Yucca Flat, Nevada, a total of eight nuclear bombs were exploded during Operation Tumbler-Snapper. However, only Shot Charlie allowed camera crews to be present. The drop site for Charlie was located five miles north of the dry lake bed. An access road stretched the entire length of the test site making for quick travel to ground zero. Furthermore, there were paratrooper drop locations north of the blast zone (for after the blast), and trench positions south of the drop zone where the troops and observers were stationed. Reporters and camera crews were located roughly one mile behind the trench positions.[14]

Along with the Tumbler-Snapper detonations the DOD also conducted a series of military exercises known as the Desert Rock Exercises. During Tumbler-Snapper the government was on Exercise Desert Rock IV.[15] This exercise mandated tactical maneuvers of both men on the ground as well as paratroopers jumping from aircraft after the explosion. Soldiers walked within 200 meters of ground zero examining the bomb site as well as military equipment left for study. Soldiers were then subject to a series of questions and psychological examinations to determine the effects of witnessing a nuclear bomb.[16] Approximately 7,350 soldiers and other members of the Department of Defense (DOD) participated in Exercise Desert Rock IV during Operation Tumbler-Snapper.[17]

On April 1 and 15, the first two bombs of Operation Tumbler-Snapper detonated. These Shots, Able and Baker respectively, were relatively low yield explosions clocked at one kiloton of energy and detonated at a height of roughly one thousand feet above ground.[18] Both bombs were airdropped with no follow-up military exercises after the explosion. These Shots provided explosion data used to calculate the power, range, and other factors of future nuclear bombs, including Shot Charlie.[19] With these tests conducted, a larger explosion (Charlie) could have better predicted results.

SHOT CHARLIE

Shot Charlie's big day came to Yucca Flat on April 22, 1952. Soldiers, government officials, military officers, reporters, and TV crews all gathered to view the spectacle. Charlie was set to drop at 9:30 a.m. Like the two detonations before, Charlie was to be airdropped from a bomber plane. However, unlike the previous bombs, Charlie would have a much larger yield of thirty-one kilotons of energy.[20] This was a substantial increase in power, even compared to the bombs of Fat Man and Little Boy (the bombs dropped on Japanese cities Nagasaki and Hiroshima), which clocked at twenty-one and fifteen kilotons respectively.[21] The larger bomb size was sure to make quite the spectacle for the spectators standing by.

The weather proved to be acceptable for test conditions, and the bomber took off. The first line of troops and observers were stationed three and a half miles away from the center of the blast. Around 1,400 personnel, primarily consisting of the troops set to drill in the blast zone after the detonation and the "all clear" call, were huddled in a meager five-foot trench. Roughly one mile behind their location were camera crews and all manner of spectators waiting for the blast. As the plane flew overhead, all fell quiet. Camera operators zoomed in on the intended blast zone and waited for the spectacle of their lifetime. The countdown began and seconds before zero, a series of flares were sent up from the ground, meant to measure the detonation height as well as the resulting shock wave. As the flares flew into the sky, their light was instantly consumed by the explosion of Charlie. The bomb mostly hit its mark, detonating only fifty-three feet below the desired height of 3,500 feet, well within the expected standard deviation.[22] Spectators on the ground, still wearing protective eye wear, watched as the ball of fire formed into the familiar shape of the mushroom cloud that has become synonymous with nuclear detonation.

Screenshot of government film capturing the dog test of Operation Tumbler-Snapper. Source: https://commons.wikimedia.org/wiki/File:Tumbler-Snapper_Dog.gif. Public Domain.

The picture above shows the bomb immediately after the detonation. The smoke trails visible on the left side of both images shown are those of the test flares. Their trails have stretched out due to the shock wave having passed through them. Furthermore, these images show the stunning display of light produced by the bomb. The bomb's brightness approximated the brilliance of the sun. Despite the awesome power demonstrated by the light show, the bomber that dropped the bomb flew away unharmed, and the ground troops arose from their trenches and gazed into the ball of fire. The detonation was conducted without incident, and officials started collecting and analyzing the data.

After the star-like brightness faded the spectators removed their goggles, and reporters began working, gathering statements from the other civilians around them. Grant Holkom representing KTLA News interviewed several onsite witnesses and collected their opinions. Grant interviewed Hugh Baillie, president of the United Press, who stated, "I have seen a lot of bombardments but nothing like that… it was of course, one of the most tremendous things the eye of man can witness, and if there were a city out there now, other than a few smoke plumes here and there it would be devastated, filled with dead people. Not many dying most of them would be dead I expect."[23] Baillie was a seasoned war correspondent, who was nevertheless impressed by the power of Charlie. His testimony suggests how impactful the bomb was on those who experienced it in person.

Later in the interview Baillie described the shock and heat waves of the explosion. The shock wave was "a combination of someone giving you a shove and slap".[24] The shock wave of a nuclear blast of this size and distance could have easily knocked someone over. Baillie also described the heatwave that followed the blast, saying "it was like opening the door of a furnace."[25] Other spectators noticed the amount of dust and sand moved by the blast. Reuters reported that "the vacuum created by the blast sucks up the sand like a cyclone."[26] Baillie's opinion was shared by almost all present at the test site.

William L. Laurence, a reporter with arguably the most nuclear experience, was also interviewed by Grant Holkom. Laurence had been the only reporter present at the Trinity bomb test—the first nuclear bomb test—and served as the official historian of the test.[27] Laurence had also been on the Great Artiste, an observation plane that accompanied the B-29 Superfortress aircrafts that dropped the bomb over Hiroshima and Nagasaki in 1945, though Laurence was only present during the bombing of Nagasaki.[28] His exclusive opportunities allowed him to write exclusive articles for various publications of nuclear weapons. After witnessing Charlie's detonation,

Laurence stated, "My reaction Grant, is really that you never see a second atomic blast… it's always a first."[29] His subsequent statements and overall account discussed the science behind the bomb, such as explaining the size of uranium core, the chain reaction caused by splitting neutrons off other atoms, and the general power stored in the atom.

An hour after the detonation, radiological tests concluded it was safe to conduct the military drills under Exercise Desert Rock IV. Paratroopers were dropped north of the blast zone and conducted maneuvers inside the blast zone after the radiation levels were safe. The troops in the trenches, as well as the observers stationed with them, leapt out and marched in units around the devastated debris field. Before the explosion the soldiers had been concerned of their proximity to the detonation. Some of them even feared they might lose the ability to have children.[30] After the explosion and subsequent march many of their safety concerns dissipated, though they had seen the destruction brought by the bomb and fell in awe of its power.[31]

Those present at the blast site saw the bomb as a creation to rival the power of God and should be feared as such. Many shared the sentiments of Laurence, Baillie, and the troops participating in the event. However, their opinions were not the only ones to be considered. In fact, their opinions of the bomb paled in comparison to the millions of people watching the bomb on their television set. The watchers on the other side of the screen did not share the same sentiments as those present at the bomb site.

WATCHING THE BOMB AT HOME

A question that the reporters could not ask April 22 was "what did the people watching from home think of the blast?" In the following days that question was answered as people across the country turned to newspapers to voice their opinion of the bomb and its telecast. Headlines about the bomb started appearing in hundreds of different publications. However, these headlines did not preach the same "power of God" sentiment felt by the people at the bomb site. Instead, they described a lackluster and unimpactful broadcast. The most common sounding headline followed the lines of this piece in the Chicago Daily Tribune, "Atomic boom on video turns into 'dud' here: Transmission difficulty hampers reception".[32]

Soon opinion pieces started to appear after the blast criticizing the dangers of nuclear bombs. One article that appeared in the Los Angeles Sentinel titled "Blast the Atom Apathy" reported, "Not only did the blast last week at Yucca Flat make for a spectacular show, but it also served as convincing proof that the great dread of radiation is largely baseless".[33] Many watchers opined that, since troops could make maneuvers in the blast zone a mere hour after detonation, radiation was no cause for concern. This belief was supported by several reports that noted troops could have safely moved forward much earlier into the blast zone.[34] Other articles like "Barrage of Publicity On Atomic Explosion Lifts Fear of Weapon: Seen on Television" argue that the atom bomb, long described in hushed fearful tones, no longer merited a fearful response. Now "something of the mystery built up around this lethal instrument has been blasted out of people's consciousness by the enormous barrage of publicity given to it by press, radio, and television."[35] Another article, this time published in the national daily The Washington Post, was titled "Disappointment, The Bomb Seems Dull on Television;" in it the author mocked the broadcast studio's claims of "technical difficulties" as an "excuse" for the lackluster picture.[36] The author summed up the reaction of another person in the room watching the screen: "Well, it went off."[37]

The explanation for viewers' disappointment is actually very simple: television was relatively new. It held the promise of allowing viewers to "see" events in real time, but the technology of "seeing at a distance" was still a work in progress. Televisions at this time were in the process of being put in every American home. In 1950, only nine percent of homes in the United States had television sets. Ten years later in 1960 that number was ninety percent.[38] The quality of live broadcasts was terrible—especially compared to our own time—even compared technology that would emerge just a few years after the explosion. Almost all early TVs had poor picture resolution, and different conditions appeared differently onscreen. The conditions of "outside broadcasts"—pictures broadcast outside the confines of the studio, made it difficult to offer the kind of images viewers might have expected from film. Broadcast equipment used to broadcast the bomb live was also in early stages of development. The first live broadcast from outside the studio had occurred less than a year earlier in September 1951.[39] The quality of these live broadcasts was much worse than regularly schedule studio-produced programs of the age.

In particular, the differences between live television and filmed programming were still very stark. For example, the picture of the bomb seen earlier in this chapter was taken from film capturing Shot Dog (the last of the Operation Tumbler Snapper tests) and reflects the brilliance of the blast, suggesting the awesome power of the bomb. However, televised images were much less representative of the blast. Instead, images that people watching live television actually saw were more like the images below.

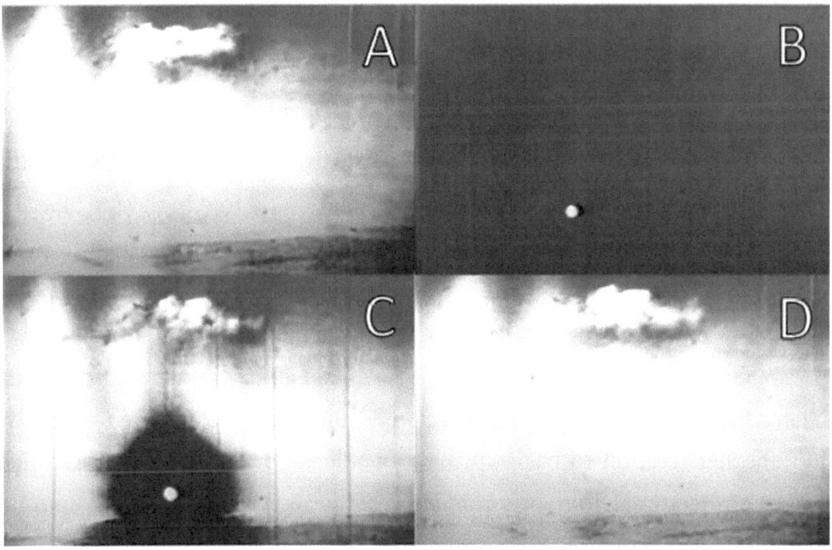

Screenshots from KTLA Live Footage: Atomic Bomb Coverage, Reported by Grant Holcomb. April 22, 1952. See original film here: https://www.youtube.com/watch?v=OF3JvVJMtzg&list=WL&index=6. Used under fair use.

The images above depict the nuclear blast as broadcasted by the KTLA news station. Picture A depicts the scene one second before the bomb blast. The resolution is very poor, and all that is decipherable is the horizon as well as a few clouds contrasted by the sun. Picture B is very dark, paradoxically showing the explosion's brightness: the light reads as darkness on the screen. The bomb center, the white dot, cast out extremely bright rays of light in which made everything around it dark, including the sun. Picture C shows the light receding and the darkness taking up less of the screen. Note that the blackness is not the mushroom cloud, but the light emitted by the blast. In picture D, the light of the bomb has completely dissipated, and if viewers look closely, they can see the some representation of the formation of the mushroom cloud. The time lapse between Image A and Image C was approximately one second. The light around the bomb in image C disappeared to form image D in six seconds. The part of the whole event that people expected to be most visually interesting took less than eight seconds. After another seven seconds elapsed, the mushroom cloud in image D was no longer visible.

One of the most surprising opinions shared by those who witnessed the bomb in person and through the screen in the own living rooms was that the atom bomb was a mercy weapon. Hugh Baillie later wrote in the *Los Angeles Times* that "One bomb does it all. The usual slow, costly infantry advances against hostile fire are dispensed with."[40] Baillie's view of nuclear bombs seemed to have shifted after contemplating the actual battlefield use of the bomb. Gladwin Hill, a writer for the *New York Times* who not present at the Yucca Flat, said "from a tactical standpoint, nuclear missiles were essentially just another weapon to back up the foot solider."[41] Hill later repeated the view of government officials that nuclear weapons were just a common tool in modern warfare, suggesting that both marines and nuclear bombs would become important units on the battlefield.[42] General Joseph Swing (present at detonation) was very pleased with the event. However, Swing desired a smaller more tactical bomb to be used in combat saying, "We must learn the size of the weapon that will show the quickest devastation on the enemy stronghold and that will also enable the infantry to move in as fast as possible."[43]

The perception of nuclear weapons as delivering "mercy" is better understood when one realizes that, during the same period as the Shot Charlie broadcast, thousands of gallons of napalm were being dropped over Korea. Napalm, a gelatinous gasoline, was considered relatively worse. Several newspapers published articles such as "Napalm Bomb Denounced" that criticized the use of napalm.[44] The fiery destruction of napalm, then overshadowed the nuclear threat, unintentionally condoning the nuclear "mercy" weapon over napalm. The public approval of nuclear bombs in war was not because of support, but because of a lack of interest and understanding of its power.

The broadcast of Shot Charlie convinced many people at home that nuclear bombs were no longer something to be feared. Those individuals who watched the detonation in person at first seem to provide the reaction our own contemporaries might associate with nuclear explosions: a panic-stricken realization of the amount of damage caused by these bombs. However, many observers soon began to believe that this weapon would save more lives than it would destroy. Even after knowing what happened to Hiroshima and Nagasaki civilians tended to see the potential "good" that came with nuclear weapons. The ultimate take away was although American citizens were aware of the Japanese bombings, they simply saw Charlie's broadcast as just another television show, overlooking the destructive power of nuclear weapons.

Notes

1. Jean Ponton, Carl Maag, Mary F. Barrett, and Robert Shepanek, "Operation Tumbler-Snapper : 1952," United States Atmospheric Nuclear Weapons Tests, Nuclear Test Personnel Review, October 1982.
2. United States. National Nuclear Security Administration. 2015. United States Nuclear Tests : July 1945 through September 1992. Las Vegas, Nevada: U.S. Department of Energy, National Nuclear Security Administration, Nevada Field Office.
3. Gene Sherman, "Gls 4 Miles from Biggest A-Bomb Blast: Record Explosion for Continental U.S. Set Off 3500 Feet Up," Los Angeles Times, Apr 23, 1952.
4. Carey Sublette, "The Soviet Nuclear Weapons Program," The NuclearWeaponsArchive.org, updated 12 December 1997. https://nuclearweaponarchive.org/Russia/Sovwpnprog.html.
5. Andrew Glass, "Truman Reveals Soviet Union is Now a Nuclear Power, Sept. 23, 1949," Politico.com. Last modified September 22, 2017. https://www.politico.com/story/2017/09/22/truman-reveals-soviet-union-is-now-a-nuclear-power-sept-23-1949-242944
6. David Fedman, and Cary Karacas, "A Cartographic Fade to Black: Mapping the Destruction of Urban Japan during World War II," Journal of Historical Geography 38. no. 3 (2012): 306-328.
7. Andrew G. Kirk, Doom Towns: The People and Landscapes of Atomic Testing: a Graphic History (New York: Oxford University Press, USA, 2017).
8. Ponton, et al., "Operation Tumbler-Snapper."
9. Ponton, et al., "Operation Tumbler-Snapper."
10. Defense Technical Information Center, V.A. Byrnes, Flash Blindness: Report to the Test Director, March 1953
11. U.S. Library of Congress, CRS, U.S. Strategic Nuclear Forces: Background, Developments, and Issues, by Amy F. Woolf, CRS Report RL33640, (Washington, DC: Office of Congressional Information and Publishing, December 14, 2021).
12. Hazel Gaudet Erskine, "The Polls: Atomic Weapons and Nuclear Energy," The Public Opinion Quarterly 27, no. 2 (1963): 155–90. http://www.jstor.org/stable/2746913.
13. Erskine, "The Polls," 1963.
14. Ponton, et al., "Operation Tumbler-Snapper."
15. Ponton, et al., "Operation Tumbler-Snapper."
16. Ponton, et al., "Operation Tumbler-Snapper."
17. Ponton, et al., "Operation Tumbler-Snapper."
18. Ponton, et al., "Operation Tumbler-Snapper."
19. United States Department of Defense, Armed Forces Special Weapons Project

(U.S.), and United States Department of Energy Albuquerque Operations Office. Military Participation on Tumbler/Snapper. [Albuquerque, N.M.]: U.S. Dept. of Energy, Albuquerque Operations Office, 1996. VHS

20. Ponton, et al., "Operation Tumbler-Snapper."
21. "Little Boy and Fat Man," National Museum for Nuclear Science and History, https://ahf.nuclearmuseum.org/ahf/history/little-boy-and-fat-man/, last modified July 23, 2014.
22. Ponton, JRB, and United States. Nuclear Test Personnel Review.
23. Grant Holcomb, KTLA News, "Atomic Bomb Coverage," News Report, April 22, 1952. https://www.youtube.com/watch?v=OF3JvVJMtzg&list=WL&index=6
24. Holcomb, "Atomic Bomb Coverage."
25. Holcomb, "Atomic Bomb Coverage."
26. Reuters, "USA: Atomic Bomb Test," British Pathé, 1952. https://www.britishpathe.com/asset/170881/.
27. Holcomb, "Atomic Bomb Coverage."
28. William L. Laurence, "Eyewitness Account of Atomic Bomb Over Nagasaki," Atomicarchive.com: Exploring the History, Science, and Consequences of the Atomic Bomb, last modified September 9, 1945. https://www.atomicarchive.com/resources/documents/hiroshima-nagasaki/eyewitness-nagasaki.html.
29. Holcomb, "Atomic Bomb Coverage."
30. Military Participation, Department of Energy.
31. Military Participation, Department of Energy.
32. "Atomic Boom on Video Turns Into 'Dud' Here: Transmission Difficulty Hampers Reception," Chicago Daily Tribune, Apr 23, 1952. http://login.ezproxy.lib.vt.edu/login?url=https://www.proquest.com/historical-newspapers/atomic-boom-on-video-turns-into-dud-here/docview/178315974/se-2.
33. "Blast the Atom Apathy," Los Angeles Sentinel, May 1, 1952. http://login.ezproxy.lib.vt.edu/login?url=https://www.proquest.com/historical-newspapers/blast-atom-apathy/docview/562303057/se-2.
34. Harlan Trott, "Barrage of Publicity on Atomic Explosion Lifts Fear of Weapon: Seen on Television," The Christian Science Monitor, Apr. 23, 1952. http://login.ezproxy.lib.vt.edu/login?url=https://www.proquest.com/historical-newspapers/barrage-publicity-on-atomic-explosion-lifts-fear/docview/508516714/se-2.
35. Harlan Trott, "Barrage of Publicity," 1952.
36. John Crosby, "The Bomb Seems Dull on Television," The Washington Post, Apr 29, 1952. http://login.ezproxy.lib.vt.edu/login?url=https://www.proquest.com/historical-newspapers/bomb-seems-dull-on-television/docview/152459125/se-2.
37. Crosby, "The Bomb Seems Dull."

38. "Research Guides: American Women: Resources from the Moving Image Collections: Television," Research Guides at Library of Congress, https://guides.loc.gov/american-women-moving-image/television.

39. "Truman to Be Televised In First National Hook-Up," The New York Times, September 4, 1951, 2.

40. Hugh Baillie, "Atomic Blast seen as Test of 'Mercy' Weapon," *Los Angeles Times* (1923-1995), Apr 23, 1952. http://login.ezproxy.lib.vt.edu/login?url=https://www.proquest.com/historical-newspapers/atomic-blast-seen-as-test-mercy-weapon/docview/166364368/se-2.

41. Gladwin Hill, "Atomic bomb stirs no awe in marines: Corps commander in explosion maneuver slated tomorrow sees just another weapon," *The New York Times,* Apr 29, 1952. http://login.ezproxy.lib.vt.edu/login?url=https://www.proquest.com/historical-newspapers/atomic-bomb-stirs-no-awe-marines/docview/112373327/se-2.

42. Baillie, "Atomic Blast," 1952.

43. Seymour Korman, "Giant A-BOMB blots out sun, awes viewers: troops 3 1/2 miles away jolted," *Chicago Daily Tribune*, Apr 23, 1952. http://login.ezproxy.lib.vt.edu/login?url=https://www.proquest.com/historical-newspapers/giant-bomb-blots-out-sun-awes-viewers/docview/178377808/se-2.

44. "Napalm Bomb Denounced: Archbishop of your urges use of weapon be outlawed," *New York Times*, Apr 28, 1952. http://login.ezproxy.lib.vt.edu/login?url=https://www.proquest.com/historical-newspapers/napalm-bomb-denounced/docview/112334328/se-2.

Part II: Propaganda

Another major opportunity television provided was as a new channel for propaganda. As it became more accessible around the world, television became crucial for framing global conflicts for regional and national audiences. Both in the realm of entertainment and news, producers focused on drawing ideological battlelines and ensuring their audiences knew which side they should be on. As many of these chapters will discuss, that often involved the creative use of portrayals of groups considered to be the enemy, such as the Soviet Union, to create a more heroic image of western liberalism.

Cold War Spy Television: *I Led 3 Lives* and Communism

CHASE FRAZIER

In 1953 Julius and Ethel Rosenberg were both found guilty of conspiracy to commit espionage and sentenced to death after they were found to have given the USSR details about the atomic bomb in 1945.[1] Their trial had attracted the media and the public and become a controversial and popular topic in the United States.[2] Cases like this increased not only the spotlight on espionage but also the number of Americans who were aware of some aspects of spies and their work, and at a moment when television producers were looking for popular stories. Television production companies capitalized on this trend by producing shows about spies or featuring spy-like characters. These shows did not just tell spy stories; they grappled with contemporary fears of communism and perceived communist infiltration of American society, thus utilizing these fears to relate to and gain an audience, while also perpetuating these same fears in the minds of Americans.

This chapter examines American spy television shows of the 1950s and, in particular, uses the example of *I Led 3 Lives* (1953-1956) to demonstrate how television programming drew upon and added to the growing fear of communism in early Cold War America. Spy television of the early Cold War took the growing fears of Americans and put them on the small screen, capitalizing on these fears to gain a larger audience. By using these fears, spy shows played a role in the further spread of American fears during the Cold War period.

Television in the early Cold War era has been explored in many scholarly works, many even focusing on or using *I Led 3 Lives* in their research. Michael Kackman's *Citizen Spy: Television, Espionage, and Cold War Culture* looks at spy characters on American television and how spy shows interacted with American culture. Kackman explores many spy shows and characters from the time period but goes in-depth on *I Led 3 Lives*, looking at the show's characteristics and how they related to culture and gender roles in Cold War America.[3] Wesley Britton also explores the topic in his book *Spy Television.* The book explores a broad history of spy television and discusses the propaganda-like aspects of *I Led 3 Lives* and many other earlier spy shows.[4] Both of these books are valuable sources in the understanding of the spy genre and *I Led 3 Lives*' place in the Cold War. This chapter aims to look at Cold War spy television, focusing on *I Led 3 Lives* to prove that the production of spy shows impacted the growth in fear of communism in America and that the shows used this same fear to their advantage.

Fear of communism became the driving force behind the United States' efforts against the Soviet Union over the forty-five year period of the Cold War. Such fears emerged first during the Bolshevik Revolution (1917), grew in the 1920s, reached their highest point during the late 1940s and 1950s, and persisted through the Cold War and onward.[5] The World War II alliance between the US and Soviet Union was short-lived, whereafter the differences between the Soviets and Americans in their respective social, political, and economic systems escalated into the Cold War and a nuclear arms race.[6] As the Soviet Union gained more power and influence after winning World War II and emerged alongside the United States as a world superpower, the second "Red Scare" in America grew with it.[7] McCarthyism is commonly considered to be the second "Red Scare," with the first occurring around the time of the Bolshevik Revolution of 1917. After World War II, the Soviet's goal of spreading communism and their sphere of influence throughout Eastern Europe was met with United States opposition in the form of the Marshall Plan.[8] This initiative allowed the United States to provide supplies and other aid for European countries that were thought to be under threat of Communist takeover. These opposing actions, along with the already rivaling ideologies of the two countries, brought them into the Cold War, cemented the two countries' friction in the early years of the conflict, and created the landscape for many of the fears Americans had throughout the period.

The fear of communism pervaded both foreign policy and domestic politics in the United States during this era. Anti-communism impacted the daily life of Americans, creating an atmosphere of suspicion among citizens. Some in power, such as Senator Joseph McCarthy and Director of the FBI J. Edgar Hoover, grew concerned with what they saw as the growing influence of the Soviet Union and worried about the loyalty of those in government positions. Senator McCarthy was a driving force in the fear of communism during this period and inspired the previously mentioned term "McCarthyism."[9] Officials under Hoover and McCarthy investigated the "infiltration" of Soviet ideas and agents within the United States government and otherwise sought to limit access to anyone perceived as untrustworthy. This was done by implementing systems such as background checks to ensure officials were not secretly communist or under the influence of the Soviet Union, for example.[10] These changes accelerated the growing fear of spying and infiltration in the United States and brought the ideas and fears of espionage in front of the American public.

Spying played a significant role in the fight against communism in America during the Cold War, as it has in American politics since the Revolutionary War. An early example of American espionage can be found in the experience of James Armistead Lafayette, an enslaved spy whose work was crucial to the American victory at Yorktown, demonstrating the importance of the practice of spying even before the founding of the nation.[11] American espionage reached new levels during the Cold War: intelligence operatives working for the United States fed information about the Soviet Union and the countries within its sphere of influence throughout the Cold War.[12] This history of spying in America showcases how ingrained it is into the country's operations. However, the United States was not the only country that utilized the tool of espionage during the Cold War. The Soviet Union had its own espionage system, spying on the United States and its allies throughout the war.[13] The Soviet Union had much to gain from getting information and intelligence from America, such as US plans for halting the spread of communism and information on the creation of the atomic bomb.[14] Orchestrating an espionage operation in the United States allowed the Soviets to catch up to the United States and test their own atomic bomb in 1949, only four years after the bombings of Hiroshima and Nagasaki in 1945.[15]

THE SPY GENRE

Spy television shows emerged during the 1950s and continued to grow in popularity through the 1950s and 1960s, with many of the most popular and well-known shows running through the 1960s. Shows, such as *The Man From U.N.C.L.E* and *I Spy,* airing in the mid-1960s played a significant role in mainstreaming the spy show genre by further popularizing the genre and finding commercial and critical success.[16] These shows continued to explore themes and ideas presented by earlier shows like *I Led 3 Lives*, illustrating the enduring effect earlier shows had on the genre.

Spy shows have many defining characteristics that differentiate them from other genres and are present in most of the depictions of the genre. The central theme of the shows is the use of secret agents or spies. Many of the spy shows in the 1950s and 1960s found themselves in the genre of spy drama, though some branched out and borrowed themes from other genres in combination with spy themes. The shows typically showcase highly trained, skilled agents on espionage missions. Another central theme is the use of technology to aid these agents; whether it be a hidden camera or microphone, gadgets are an established part of the spy genre.[17] Many spy shows, especially examples from the Cold War era, utilize the tensions between the United States and the Soviet Union as the primary conflict for their stories.[18] Another important theme commonly found in spy shows is the use of loyalty or betrayal to create conflict and plot in the story. These loyalty and betrayal plot points can be used to connect real-world espionage to these shows.[19] The shows used this idea to showcase how loyalty to one's country is an important aspect of what makes a spy useful for their country. These are some of the major themes of spy television that allowed its popularity to grow through the 1950s and 1960s, utilizing many themes from other genres to solidify itself into the mainstream of American television.

I Led 3 Lives, and the spy television genre in general, were aided by other forms of spy media that helped popularize and shape the genre. Spy television held its own in the airspace of Cold War American television, and many contributing factors enabled the creation and popularization of the genre. One of the first factors that played a part in the rise of this popularity was the increase in spy-related books being published and finding success during the early years of the Cold War. *I Led 3 Lives*, based on real-life American spy Herbert Philbrick's novel, is an example of this process in action, using the newer technology of television to give audiences a visual component to the stories told in Philbrick's book. On top of non-fiction books adapted for television, novels about spies became popular during these years, leading to the development of more television shows to capitalize on the trend. Ian Fleming, one of the more notable examples of authors who created spy fiction novels through the early Cold War, wrote many fictional spy stories, including *Casino Royale* in 1953, the first novel in the James Bond series.[20] Books played a significant role in the recognition of the spy genre and directly inspired the creation of many influential shows, such as *I Led 3 Lives,* through both the 1950s and 1960s.

The James Bond book series would also get a successful series of movies starting in 1962 that significantly boosted the popularity of the spy genre in America and jumpstarted the production of many shows during the 1960s.

The first of the James Bond movies, *Dr. No*, was released to great reception and success in 1962, making more than fifty million dollars at the box office worldwide.[21] This success popularized the James Bond character, leading to five follow-up films in the 1960s and many more in subsequent decades. The success also helped shift audience views on spies and espionage through the peculiarities of James Bond's character, depicted as a sophisticated and professional agent.[22] Unlike Herbert Philbrick, the protagonist of *I Led 3 Lives*, Bond was a fictional spy based heavily on the themes of real-life agents, allowing Fleming and the James Bond films to create the ideal spy for entertaining audiences while also keeping aspects of real espionage.[23] The films set a precedent for how media would depict spies while being one of the major contributing factors to the success of the spy genre that allowed for the wave of 1960s spy shows.

Evolution of American spy shows with other genres in the 1960s kept the genre alive even after extremism of the McCarthyism era, as heightened fears of communism diminished as a driving force behind the production of these shows. This evolution included the branching out of spy shows into other popular categories.[24] Many still used the themes of drama, one of the major sections of television that is often combined with other themes and genres. The drama genre has specific characteristics, including a focus on developing characters, a longer and more detailed story, and an exploration of emotions and consequences in serious situations.[25] Spy shows also explored many other genres that were commonly combined with drama series at the time, intertwining the spy drama genre with other popular themes. Shows like *I Spy* and *Get Smart* mixed comedy and spy drama, branching the spy characters and themes into a more lighthearted genre.[26] This genre diversity allowed the shows to connect with different audiences, leading to an increased viewership of the genre as a whole.

Outside of books and movies, the spy genre capitalized on real-world events and news. The contentious relationship between the Soviet Union and the United States gripped public attention, and the tensions of the Cold War were a part of everyday life for Americans living through this time.[27] One of the most well-known examples of spying that came to the public eye was the trial of Julius and Ethel Rosenberg, as previously mentioned. The American couple's public trial, which ended in their execution in 1953, brought mass media attention that bolstered the public's interest in and knowledge of spies.[28] This event, along with others, brought spying into a more public space and allowed for many forms of media surrounding the topic of espionage to be produced and find success.

Spy shows used the tensions of the Cold War as plot devices and settings for their characters, taking advantage of the popularity of spy books and movies through the early Cold War period. The Red Scare was a significant contributor to Americans' fears during the early parts of the Cold War, and spy shows both added to and utilized this fear of communism. Spy shows allowed audiences to relate their fears to something on television, while also being entertained by the fight against communism, showcasing an often fictitious version of how the communist threat was being fought in America.

I LED 3 LIVES

Before the flood of spy programming came to American television in the 1960s, one show pioneered the theme of espionage on the small screen in the early 1950s: *I Led 3* Lives (1953-1956, NBC).[29] As one of the first spy shows on American television, *I Led 3 Lives* laid the groundwork for many of the more popular shows that came later in the 1960s. The show introduced many of the genre's themes and conventions to American audiences while helping to establish them in television. *I Led 3 Lives* provides an interesting case because, unlike many of the spy shows from the era, its plot lines stem from the real life of Herbert Philbrick. An American citizen, Philbrick became a member of the Communist Party in America in the early 1940s, before becoming a FBI informant and reporting on the activities of American communists. He worked with the FBI through most of the 1940s as a counter-espionage agent, infiltrating various communist groups nationwide. His work put him in danger and required him to join many communist organizations, attend their meetings, and report on the actions of communist groups and individuals, all the while keeping up his real-life responsibilities.[30] Philbrick finally broke his cover to testify against a group of American Communist Party leaders that he had reported to the FBI. After a successful conviction of these American Communist leaders, he ended his operation with the FBI in 1949.[31]

Philbrick wrote a book about his work with the FBI and his dealings with the Communist Party in America, titled *I Led Three Lives: Citizen, 'Communist,' Counterspy* (1952).[32] The book became a best-seller in the United States, leading to its eventual adaptation for television in 1953. The show delves into the life of Herbert Philbrick, showcasing the three lives he lived, as an ordinary citizen, as a member of the Communist Party, and as a spy for the Federal Bureau of Investigation (FBI). While the show was inspired by Philbrick's life, only the show's first two seasons are based directly on the book. The third season is based on some of his files and the experiences of other agents after the show exhausted the book's content. Each episode of the series began with the narrator and star of the show, Richard Carlson, introducing the show to the audience and explaining that the story they were about to watch was based on fact. The show focuses on the stress that Philbrick had while trying to lead these three opposing lives, taking the audience through his thoughts while traversing the difficulties of hiding each of his identities.[33] Overall, the show played a significant role in introducing elements of the spy drama genre to American audiences and laid the groundwork for future spy shows while having its own impact on its audience.

I Led 3 Lives had an influential role in creating and popularizing future spy media, but it was also successful in its own right. The show ran for three seasons, airing 117 episodes over its three-year run.[34] *I Led 3 Lives* did well in television ratings, one of the main factors when looking at the success of television shows in the network era. *Broadcasting Magazine* reported in August of 1955 that the show had reached the top ten shows in a majority of the major television markets in the United States, for example.[35] The show also received critical recognition, earning two Primetime Emmy nominations in 1954 and 1955.[36] These achievements show the popular and critical success of *I Led 3 Lives* and indicate the marketability of the spy genre in television, leading to the development of future spy shows.

In the show *I Led 3 Lives,* the main threat of communism comes from domestic communist parties and groups infiltrated by spies like Herbert Philbrick. Although it was common for spy shows to utilize foreign Soviet spies or foreign communist threats as the antagonists in their stories, *I Led 3 Lives* establishes an example of a home-grown threat to Americans. The show portrayed communist Americans participating in the Communist Party and working toward dismantling the United States as a capitalist country. In the show, communists are portrayed as secretive and hostile to the freedoms and liberties of the United States as a whole; they serve as the antithesis to the main character. The communist parties throughout the show are shown to be ruthless groups that are plotting to overthrow the government, even discussing the arming of citizens for a forceful overthrow of the government in the very first episode of the series.[37] Herbert Philbrick, the protagonist and narrator of the series, fights heroically against homegrown communism and keeps America safe. He is a positive role model built to appeal to the contemporary American audience. While many of the communist characters in the show are shown as the opposite, Philbrick offers a character that suggests the "correct" ways of life to American viewers by showcasing anti-communist beliefs and actions while also living a secret life as a spy for the FBI and one as an ordinary American citizen. These depictions of the characters allow the show to display their intended narrative of how dangerous the Communist Party was, how important the fight against communism was, and how important American citizens' role was in that fight.

I Led 3 Lives along with spy shows from this era utilized the fears held by Americans and lined up with how government officials such as Senator McCarthy portrayed communists and communism.[38] Communists were the main antagonists in every episode and were consistently depicted as a danger to everyday Americans. The communists shown in *I Led 3 Lives* were portrayed as sneaky and as wanting to infiltrate America disguised as ordinary citizens or officials. This lines up with many of the things Americans were hearing from people like Senator McCarthy, who stressed the idea that communists had infiltrated the American government. One notable example of this is a speech he gave in 1950 in which he claimed to have a list of 205 people that worked for the United States government that were known to be members of the Communist Party. The media uncritically covered McCarthy, spreading his claims and instilling fear in Americans.[39] By playing on the fears that McCarthy helped spread, spy shows dramatized current events to their audiences and more effectively influence the opinions of Americans.

Spy television shows of the early Cold War clearly used the geopolitical situation and fear of communism to their benefit regarding their message and popularity. However, these shows also aided in reinforcing and spreading the fear of communism in America. One of the ways the shows did this was by portraying communist agents and members of the communist groups in America as ruthless enemies of the United States with the primary goal of destroying the systems and daily lives of Americans. Viewers of the show who already harbored their own fears and beliefs about communism felt validated by the portrayal of communists in spy shows. Communists were portrayed to relate to the most common fears of Americans, such as being shown directly threatening the safety and way of life of everyday Americans.[40] The shows also had the ability to instill fear in those who may not have known much about communism and what it was about and constantly showcased the dangers of communism to those who may have been skeptical of the idea that it was a threat.[41] Many of the shows, including *I Led 3 Lives*, did not often or accurately explain what communism even was, only displaying communism as the enemy and implying it was bad.[42] This lack of explanation was an important aspect of the show's tactics for dealing with the subject of communism. *I Led 3 Lives* centered around communism and its threats while often leaving viewers to use their own predetermined or learned ideas about communism to explain the actions in the show. By not delving too deeply into the tenets and beliefs of American communists, the show spread the fear of communism without educating its audience on exactly why it should be considered a threat. The shows played a role in the creation of a more anxiety and paranoia-filled country, leading to a more vigilant populace regarding communism and its perceived threats.[43] Spy shows also defined the central conflict as the geopolitical rivalry between the United States and the Soviet Union, the most important aspect of which was portrayed as the spread of communism in the world and at home.

Shows such as *I Led 3 Lives* used the idea that the United States stood against the Soviet Union to encourage the responsibility of all American citizens to help stop the communist threat. The show televised stories of everyday citizens making a difference in the face of the communist threat. In one episode of the show, for example, a mother suspects her son of joining the Communist Party. "Commie Son," depicts a Communist Party that is persuasive in recruiting members—especially youths—to their side, poses a danger to the lives of Americans, and encourages callous behavior, even towards their own members along with their families. The main plot point of the episode revolves around the mother contemplating whether or not to report her son to the FBI for associating with communists. She asks Philbrick, her neighbor, for advice. He encourages her to report her son to the FBI. In the end, however, the son turns on the party due to Philbrick's influence and rejoins the capitalist world.[44] The episode's narrative dilemma explores the idea of turning a close friend or family member into the FBI for suspicion of communist activity. The show explores many similar moral dilemmas, and most end on a high note, with Philbrick and other characters outmaneuvering the communists and aiding the FBI, sometimes leading to characters abandoning the party and returning to "normal." This type of storytelling heavily encouraged reporting any perceived communist activity by Americans. The show did not, however, represent what might happen to those accused of deviation by well-meaning relatives or friends. Such reports often led to punishments for those who had flirted with communism.[45] The dramatization of these scenarios modeled for American audiences the "proper" way to deal with communism in their own lives and communities.

The influence of spy shows such as *I Led Three Lives* not only shaped behavior, but also "hearts and minds," helping to bolster the reputation of the FBI. Both Philbrick's book and the series portrayed the FBI in a positive light, displaying the organization as the primary defense the United States had against communism at home.[46] When Philbrick testified against the group of communists in 1949, he broke his cover, meaning that he could no longer safely infiltrate communist groups and turn over information to the FBI.[47] But telling his story allowed him to continue his fight against communism in a new way. The show helped promote the image of the FBI as a productive and effective part of the government.[48] While the FBI did not directly fund the show or put its name on it, and there is no evidence that the agency played a role in any aspect of the show's production outside of Philbrick's involvement, it did have a lot to gain from positive media portrayals, like *I Led 3 Lives*. The show provided an opportunity to propagandize ideas about communism and to recruit ordinary Americans into the fight against it.[49] The show was apparently a favorite of J. Edgar Hoover, the director of the FBI, who wanted viewers to report their suspicions of anyone affiliated with communism.[50] The FBI was an essential part of the fight against communism in *I Led 3 Lives,* and the show played a role in creating a successful image for the agency.

Spy television shows like *I Led 3 Lives* both profited from and further perpetuated the fear of communism in the US during the early Cold War. The common themes and plot points of the genre spoke to the political situations and operations that occurred during the early parts of the Cold War. *I Led 3 Lives* was one of the first shows to introduce American audiences to the tenets of Cold War anti-communism through spy television in the 1950s, setting up the genre for the 1960s, when many more spy shows were broadcasted. At the same time that *I Led 3 Lives* was airing on American television, Ian Fleming's James Bond book series became popular, showing audiences the ideal fictional spy and leading to the film series in the spy-filled 1960s. Herbert Philbrick continued his work of fighting communism through the publication of his own book *I Led Three*

Lives: Citizen, 'Communist,' Counterspy in 1952, the success of which led to the production of the television show. The geopolitical tensions between the United States and the Soviet Union acted as the perfect setting for the spy genre. By exploring the anxieties of the Red Scare, spy shows connected with more potential audiences whose lives were consumed by the fear and threat of communism. *I Led 3 Lives* gave audiences a view of the potential threat of communism from the inside, showing the Communist Party in America as an organization in the shadows, plotting against the United States as a whole. The show and other shows in the genre showcased the dangers of communism and the Communist Party, using their influence to increase the fear and opposition Americans had of them. In the case of *I Led 3 Lives*, the former counterspy for the FBI, Herbert Philbrick, aided in the creation of a show that painted the FBI in a positive light and as a frontline defense against communism that could be trusted by the American people. Examining spy television through the lens of its impact and reliance on the Cold War and the communist threat serves as a useful tool to look into the impact the genre had on the minds of Americans.

Notes

1. Norman S. Beier and Leonard B. Sand, "The Rosenberg Case: History and Hysteria," *American Bar Association Journal* 40, no. 12 (1954): 1048. http://www.jstor.org/stable/25719022.
2. Beier and Sand, "The Rosenberg Case."
3. Michael Kackman, "I Led 3 Lives and the Agent of History," in *Citizen Spy: Television, Espionage, and Cold War Culture*, (Minneapolis: University of Minnesota Press, 2005). http://www.jstor.org/stable/10.5749/j.ctttshqj.7.
4. Wesley A. Britton, *Spy Television* (Westport, CT: Praeger, 2004).
5. Nick Fischer, "Here Come the Bolsheviks!: The Russian Revolution and the Red Scare," *Spider Web: The Birth of American Anticommunism* (Chicago: University of Illinois Press, 2016), 51–52. http://www.jstor.org/stable/10.5406/j.ctt18j8x4m.8.
6. Library of Congress, "Revelations from the Russian Archives: the Soviet Union and the United States," June 15, 1992, https://www.loc.gov/exhibits/archives/sovi.html.
7. Don E. Carleton, "'McCarthyism Was More Than McCarthy': Documenting The Red Scare At The State And Local Level," *The Midwestern Archivist* 12, no. 1 (1987): 13, http://www.jstor.org/stable/41101686.
8. Alexander D. Weissman, "Pivotal Politics—The Marshall Plan: A Turning Point in Foreign Aid and the Struggle for Democracy," *The History Teacher* 47, no. 1 (2013): 111. http://www.jstor.org/stable/43264189.
9. *McCarthyism* is the term used to describe the fear tactics that led to the rise of the fear of communism during the early Cold War. Carleton, "'McCarthyism Was More Than McCarthy'," 13-14.
10. Kenneth O'Reilly, "The FBI and the Origins of McCarthyism," *The Historian* 45, no. 3 (1983): 375. http://www.jstor.org/stable/24445173.
11. Kenneth A. Daigler, "Yorktown and the Endgame," in *Spies, Patriots, and Traitors: American Intelligence in the Revolutionary War* (Washington: Georgetown University Press, 2014), 226–227. http://www.jstor.org/stable/j.ctt6wpkz8.16.
12. Michael J. Sulick, *Spying in America: Espionage from the Revolutionary War to the Dawn of the Cold War* (Washington: Georgetown University Press, 2012), 265. http://www.jstor.org/stable/j.ctt3fgtvh.40.
13. Sulick, *American Spies*, 25–26.
14. Charles Ziegler, "Waiting for Joe-1: Decisions Leading to the Detection of Russia's First Atomic Bomb Test," *Social Studies of Science* 18, no. 2 (1988): 197. http://www.jstor.org/stable/285297.

15. Ziegler, "Waiting for Joe-1," 197.

16. Kackman, *Citizen Spy*, xxi.

17. Marcus M. Payk, "The Enemy Within: (De)Dramatizing the Cold War in U.S. and West German Spy TV from the 1960s," in *Cold War Cultures: Perspectives on Eastern and Western European Societies*, edited by Marcus M. Payk, Annette Vowinckel, and Thomas Lindenberger (New York: Berghahn Books, 2012), 96. http://www.jstor.org/stable/j.ctt9qdfsw.8.

18. Payk, "The Enemy Within," 105.

19. Kackman, *Citizen Spy*, 100.

20. Ian Fleming, *Casino Royale* (London: Jonathan Cape, 1953).

21. Llewella Chapman, "'They Wanted a Bigger, More Ambitious Film': Film Finances and the American 'Runaways' That Ran Away," *Journal of British Cinema and Television* 18, no. 2 (2021): 176–97, https://doi.org/10.3366/jbctv.2021.0565.

22. Drew Moniot, "James Bond and America in the Sixties: An Investigation of the Formula Film in Popular Culture," *Journal of the University Film Association* 28, no. 3 (1976): 26. http://www.jstor.org/stable/20687331.

23. Drew Moniot, "James Bond and America in the Sixties," 26.

24. Kackman, *Citizen Spy*, xvii–xl.

25. Robert McKee, *Story: Substance, structure, style, and the principles of Screenwriting* (New York: HarperCollins, 2010), 80-86.

26. Britton, *Spy Television* (Westport, CT: Praeger, 2004), 9.

27. Carleton, "'McCarthyism," 13-14.

28. Beier and Sand, "The Rosenberg Case."

29. *I Led 3 Lives,* 1953-1956, Ziv Television Programs.

30. The Associated Press, "Herbert Philbrick, 78, F.B.I. Spy Who Inspired TV Series in the 50's," *The New York Times,* August 18, 1993. https://www.nytimes.com/1993/08/18/obituaries/herbert-philbrick-78-fbi-spy-who-inspired-tv-series-in-the-50-s.html.

31. The Associated Press, "Herbert Philbrick, 78, F.B.I. Spy Who Inspired TV Series in the 50's," *The New York Times,* August 18, 1993.

32. Herbert A. Philbrick, *I Led Three Lives: Citizen, "Communist," Counterspy* (New York: McGraw-Hill, 1952).

33. *I Led 3 Lives,* Season 3, 1955-1956, ZIV Television Programs.

34. *I Led 3 Lives,* 1952-1956, ZIV Television Programs.

35. "The Top Ten Films In Ten Major Markets," *Broadcasting Magazine,* September 26, 1955.

36. "I Led Three Lives," Television Academy, accessed April 2, 2023. https://www.emmys.com/shows/i-led-three-lives.

37. Eddie Davis, Donn Mullally, Herbert A. Philbrick, "Pilot," *I Led 3 Lives,* ZIV

Television Programs, 1953.

38. Stuart J. Foster, "Chapter I: The Red Scare: Origins and Impact," *Counterpoints* 87 (2000): 5-6. http://www.jstor.org/stable/42976159.

39. Foster, "The Red Scare," 6. http://www.jstor.org/stable/42976159.

40. Carleton, "'McCarthyism," 13.

41. Payk, "The Enemy Within," 96.

42. *I Led 3 Lives,* 1952-1956, ZIV Television Programs.

43. Kackman, *Citizen Spy*, xxxix.

44. Jack Herzberg, Stuart Jerome, Herbert A. Philbrick, "Commie Son," *I Led 3 Lives,* ZIV Television Programs, 1953.

45. *I Led 3 Lives,* 1952-1956, ZIV Television Programs.

46. *I Led 3 Lives,* 1952-1956, ZIV Television Programs. Philbrick, *I Led Three Lives.*

47. The Associated Press, "Herbert Philbrick, 78, F.B.I. Spy Who Inspired TV Series in the 50's," *The New York Times,* August 18, 1993.

48. *I Led 3 Lives,* 1952-1956, ZIV Television Programs.

49. Philbrick, *I Led Three Lives.*

50. Britton, *Spy Television*, 23.

Broadcasting War: Vietnam in America

DANIELLE SCHUMACHER

As the United States battled against communism in the post World War II era, US leaders increasingly escalated the military conflict in Vietnam. By the mid-1960s many Americans began to question the purpose of the United States' involvement in Vietnam, which led to public unrest. The news media portrayed the actual events in Vietnam with reporters on the ground and, with this, demonstrated the power of the news media to shape public opinion and to fuel unrest among American citizens. By viewing the war in real time, the American public became enthralled with the Vietnam War. With the emergence of a televised war, Americans relied on information about the war from the television, and evidence began to cast doubts on the official narrative espoused by policymakers. Reflecting on the Vietnam war, veteran reporter William M. Hammond suggested a link between the media reporting and the loss of public opinion on the war, asserting that "the war in Vietnam was lost on the propaganda front."[1] The war became a prime opportunity for television to bring the news from abroad home to America, and allowed the American people to witness the unfolding events of the war. I argue that the news helped push public opinion from the days of uncritical patriotic support in World War II toward opposition. In turn, television helped propel the antiwar movement and change American perception of the war as well as government officials. For more than eight years, the televising of the Vietnam War led to the decline of American support for the war. Still to this day, most Americans believe that US military troops should not have been involved in Vietnam. The war was projected on television in a way that the American government did not anticipate, shifting public thinking of Americans. Television *news* amplified discontent with the war, while television *advertising* helped the antiwar movement gain momentum, fueling unrest and sowing distrust towards the US government. Antiwar advertisements on television framed the war in such a specific manner that escalated the antiwar movement and brought with significant changes in American military policies.

TELEVISION WAR

On-site coverage of the war became an integral part of shaping Americans' opinions on the war. The goal of the news coverage and the actual effects were incompatible in that the intended efforts were to show the high morale and patriotic efforts of America, but instead Americans saw a bleak view of the war. The realistic footage enraged many Americans as they viewed the footage of the "help" the United States contributed to in Vietnam. Americans could now see the perils of the battlefield and thus shifted their public opinion on war and American military policies. This shift in public opinion called for a new arrangement between the American people and the American government, with Americans focusing now on news networks and trusting those in charge of these networks versus that of politicians.

Previous scholars have stated that the war in Vietnam had a significant impact on everyday Americans in that the trust in the United States government shifted to the trust of the United States media shown on television. "Certainly the Vietnam War was a highly salient political event to the American public nightly television news, bringing the fighting in Vietnam and the concomitant domestic protests into our living rooms."[2] Americans now had access to direct information on the war that often contradicted the official claims of the US government.

1968 was a pivotal moment in the war for hearts and minds. "The Tet offensive in the Vietnam War, the presidential election campaign, the assassinations of Martin Luther King, Junior, and Robert F Kennedy, and the Soviet invasion of Czechoslovakia were television big stories, each producing memorable images and moments of high drama, assuring them prominent places on newscasts and the visual medium of television by 1968 more than half of the American people relied on television as their principal source of news."[3]

As getting the news became an everyday occurrence in most lives of the American people, so did owning a television. What is already known is that the media coverage of the war in Vietnam shocked Americans and brought down public opinion drastically. By 1971, a majority of Americans wanted the war to end. Partly because the war was televised, "Americans felt that the involvement up to that point had been a mistake."[4] The antiwar movement brought along a decrease in public opinion on presidents like Richard Nixon, increasing distrust between the American people and the United States government.

The Vietnam War protests produced much mass media compliance with it, such as antiwar ads, commercials, posters, news, articles, etc. that all pushed the rhetoric that America should not be involved with this war, and many A believed the war was unjust, and they could trust the media outlets to provide them with truthful information over the official claims of US leaders.

THE VIETNAM WAR

The US military intervention in Vietnam between 1955 to 1975 was controversial and caused strife between the American people and the American government. The trust between these two parties broke down as the battle against communism in Vietnam went on. By the early 1960s, television technology and cutting-edge reporting of worldwide events created the emergence of the first televised war in Vietnam. The ability to "see" the war shaped US public opinion on the war and its efforts. Televised tragedies such as the Tet Offensive further escalated the antiwar movement. I aim to examine how television framed the issue of the Vietnam War and its role in shaping public opinion of both the United States government and the presidents that oversaw the conduct of the Vietnam War. This research will include the relationship between the American people and the United States government to show how viewing the live footage of the war and antiwar advertisements shaped public opinion. This chapter will add to the historiography on the topic by focusing on the relationship between public opinion and televised war coverage.

More than 58,000 Americans died in Vietnam during one of the most unpopular wars of the twentieth century. When military troops deployed to Vietnam in 1965, most Americans hoped for a quick end to the war. After the escalation of war under under President Lyndon B. Johnson, public unrest intensified as the number of US casualties began to rise. Largely in resonded to public outcry against the death tolls, President Richard Nixon advocated for Vietnamization, a policy of withdrawing US combat troops and providing increased aid, arms, and training the South Vietnamese military. In 1970, however, Nixon expanded the war and violated Cambodian neutrality. The invasion of Cambodia generated a new wave of antiwar protests at college campuses across the United States. National news coverage of these events propelled the antiwar movement, and, in turn, the public became increasingly disillusioned.

ESCALATION OF THE WAR AND THE TET OFFENSIVE

The war generated a large antiwar movement, which has become a distinctive memory of the Vietnam War. Along with the antiwar protests, propaganda and news coverage influenced public opinion on the war and the US government. Beginning in January 1968, the Tet Offensive involved a series of North Vietnamese attacks on more than one hundred cities and outposts in South Vietnam. The news media transmitted shocking images into the living rooms of Americans, making it clear that the war was far from over, with no end in sight for Americans. This contradicted the claims that General Westmoreland bad been making. News coverage of the Tet

Offensive was one of the first in a series of tragedies caught on camera in real life, not produced in a Hollywood film studio. Graphic images brought with them widespread unrest among American citizens. The Tet Offensive was a turning point in the war for both the war in Vietnam and the war over public opinion. In some ways, this was a war for the hearts and minds of Americans—to decide whom to believe and where to put their faith, in policymakers or television screens. There were many journalists at this location filming for Americans and the world to see; "on the eve of the Tet Offensive, in January 1968 they were 179 American journalists, accredited by the military commanded Saigon"[5] Although Johnson and his administration claimed that the Tet Offensive was a military success for the United States and the South Vietnamese. What was captured on tape looked instead like a defeat. "The Tet Offensive would soon demonstrate to the American people the extent to which they had been deceived,"[6] This strategic news coverage shocked the American public and eroded much of the remaining support for the American war effort. Television changed the way that war was seen all over the world. The development of television preceded an increase in the need for creating trust between the viewer and the television from which they were getting their news. Daniel Hallin's *The "Uncensored War" The Media and Vietnam* and Michael Sullivan's *The Vietnam War A Study in the Making of American Policy* both emphasize the relationship between media and public opinion and how the televising of the war in Vietnam changed the relationship between the people and the government. "The relation between the media and government during Vietnam was in fact one of conflict,"[7] This indicated a growing tension between the media and the government. Hallin explained "The cameras were decisive in the end," creating the notion that without the cameras, the patriotic wartime sentiment of the previous wars would have continued.[8] Without cameras filming what was going on and broadcasting this to all of America who had access to a television, then there might not have been as large of a public backlash against the war. "The media coverage of Vietnam either showed literally the destructiveness and frustration of the war or-again the conventional wisdom is contradictory-presented it from a critical rather than a supportive perspective."[9] Images of destruction motivated many Americans to protest against it. As seen in the image below, Americans were seeing the mass destruction associated with the Tet Offensive and, with this, came renewed protests to bring home American troops home from Vietnam.

FILMING THE WAR

The goal of the embedded wartime coverage and its actual effects were different in that the intended efforts were to show the high morale and efforts of America, but what was shown was a bleak and shocking view of the war. The realistic footage infuriated many Americans as they watched what was happening in the places that the United States was claiming to save. These citizens could now see the perils of the battlefield that were once hidden from civilians and glamorized through past patriotic propaganda, thus shifting their public opinion on the war and American military priorities. Antiwar commercials, public approval polls, and various documents containing public opinions all provide insight into understanding the power of news coverage of the Vietnam War. What can be seen from these sources is the impact that these broadcasts had on public opinion on the war in Vietnam, along with questionable military necessity. The ability of the news to have such an impact on the American public shows the emphasis and importance of the press and mass media during a time of war.

The Vietnam War is often referred to as "the first television war" because it was the first war that was accessible to everyday Americans. From their living room, the average person could for the first time during a military campaign, view exactly what was going on across the world, creating a relationship between the media and the consumer. The war began to be televised in the early 1960s but became prominent in 1965. Televising the war unintentionally divided a nation that prided itself on unity and togetherness. Many argue that the war was lost because it was televised. Michael Mandelbaum stated that "if its previous wars had been televised, the United States would not have persevered in fighting them"[10] This counterfactual argument emphasizes the impact that televising the war had on the actual conflict itself. If the war had not been televised would it have gone as it did? Would there have been a different outcome if the war had not affected the American homefront as it did? While it's impossible to know with any certainty, when a person can examine firsthand an event in real-time, it can cause a shift in opinion or stance on that subject. If a viewer witnesses graphic content that contradicts their understanding of their nation, it can lead to feelings of betrayal and dismay. "Regular exposure to the ugly realities of battle is thought to have turned the public against the war." [11] The public wished to see the realities of the war that so many Americans were asked to fight. Once they found out through viewing in real-time, violent military tactics and defeats, then they felt increased urgency to end America's involvement in Vietnam.

The news media tended to criticize the war efforts in Vietnam and projected this onto what was televised and what Americans were going to see during the war[12] The U.S. began efforts to control the news media in the early 1960s; this would have portrayed to Americans what the U.S. government wanted it to see which would bring the two groups closer together regarding a relationship encapsulated by public opinion. The Kennedy administration struggled in 1963 with the Buddhist crisis in Vietnam, when Americans first saw on their television the horrific images of suicides and religious protests. This was one of the catalysts in the early 1960s that sparked criticism of U.S. support for the Diem government in South Vietnam. This early footage sent a wave through both the media and the public on what they should be interested in and how it affected individual people within this country. As the Kennedy administration struggled to maintain the image of the United States during this stage of intervention, the Johnson administration had similar challenges in the mid-1960s. At the height of the war, with over 600 reporters being present in Vietnam, some of whom were being captured and killed, the images struck a chord with Americans and the media. Rather than relying exclusively on information from the Johnson administration, news media broadcasted directly from the battlefields of Vietnam, offering a direct glimpse into the reality of the war from the home front. This change of perspective created a growing divergence between the U.S. government and the American public. With increasing interests, Americans watched the war violence unfold and many began to question the official justification for the war. The media portrayed the conflicts with realism and created a sense of crises, like the Tet Offensive and the My Lai Massacre in 1968 (when US troops killed 504 innocent civilians), events that contradicted what many previously believed about the war and triggered a series of riots and protests across the country in cities and on college campuses.

THE ANTIWAR MOVEMENT

As Americans saw the number of casualties rising every day in Vietnam, they became increasingly weary about the end result of this war and unsettled with the thought of remaining in Vietnam any longer. The antiwar movement originated from the shift in opinion from the American public on the Vietnam War and produced divisions between groups of people like the hawks and doves. Hawks wanted to wage the war more aggressively, and doves sought to resolve the conflict in other ways. The antiwar movement originated among groups of college students and civil rights activists in 1965, but gained national prominence and quickly spread across the U.S., including on military bases. The antiwar movement reached across America in ways that made it central to collective memory of the era. It remains an important aspect of the Vietnam War.

Americans protested the war in a number of ways, including public displays of burning draft notices and marching on the U.S. Capitol, raising larger questions about US involvement in Vietnam, and offering powerful messages that the people would rise up against government policies that were unpopular. Along with protests came a series of antiwar movement commercials that asked Americans to take a side for or against the war. These commercials questioned why the United States military was in Vietnam and tried to provoke anger and other feelings of discontent and mistrust that brought Americans a notion of wondering what their government was doing and what could be done to fight against it. With the emergence of antiwar protests in lieu of military parades, Americans began to establish a connection with the media and each other rather than with the American government. The American people found little evidence justifying American military intervention in Vietnam at this time. Antiwar campaigns and commercials flooded the news channels in the hope to persuade American citizens to oppose the war. In one example, a television commercial suggested that young boys in America would have to grow up and still be a part of Vietnam because the American military would not be pulled from Vietnam for a long time. The commercial urged Americans to write to their senators to persuade them to vote to pull troops out of Vietnam. Another commercial released on the ten-year anniversary of Vietnam featured John. F. Kerry, a former Lieutenant for the United States Navy, who said that the Vietnam War had lasted longer than any other war in American history and pitched families against families in one of the worst chapters in United States history. He cited a statistic that seventy-three percent of Americans wanted to remove troops from Vietnam and similarly urged Americans to write to their senators. The ad also criticizes the use of tax dollars for war and how much of America's money was going to be used for the Vietnam War and future wars. Both of these ads present a partisan opinion of the Vietnam War and how Americans to feel about it. Both ads appealed to Americans through the use of emotion and logic, such as preserving the lives of young men who will go to war, and the spending of tax dollars. Such ads questioned the reasons for the American troop presence in Vietnam and argued for Americans to take a stand against the war. Television advertising offered antiwar groups a new channel to challenge government policy and shape American public opinion.

THE POWER OF TELEVISION

The influence of the televising of the war allowed Americans to see what was truly going on and casted doubt on the claims of public officials. Televising the war pushed a narrative that displayed American troops completing unnecessary tasks that did more harm than good. American citizens began to distrust the American government and put their trust in the news outlets that were providing them with this information, including visuals of the war and all its sentiments. Televising the Vietnam War put a space between the media and the government, and with this came the implication that the government had lied. It became clear in the United States that one had to choose a side, and in the case of the Vietnam War and living through the 1960s-1970s, the clear choice was to side with the source that was giving raw details and images of what was actually going on with the war, or the government officials who did not provide its citizens with full accounts and the reasons why the military in Vietnam. In contrast to previous wars, the news media during the Vietnam War wanted to show the most exciting and dramatic stories with the best reporting they could have. Journalists wanted to portray to Americans a realistic glimpse into the lives of those who were on the ground fighting in Vietnam. Whether intended or not, Americans did not like what they saw of the war, and many began to trust their own eyes, rather than the government. US Marine Michael C. Mitchell has argued that "with television, we are faced with a form of technology that has the potential not only to inform but also to change the course of events"[13] With this, Mitchell warned that television wielded a new power to not just reflect but shape everyday life.

Why was television able to increasingly draw people into the war? One reason could be in part due to the power of television to simplify messages to make sense to an audience, as well as convey a journalistic tone that people take seriously. So often people take television at face value, which allows them to understand better what is going on in the world and come up with ideas of their own that they think came from television but really are exemplified by their own minds and understanding of the subject matter. Whether or not media wanted political unrest, television did foment political activism and social change.

CONCLUSION

The broadcasting of the Vietnam War shaped reality. The platform enabled news media to create and show a perception of the Vietnam War to the American public, causing an uproar about the conditions and reasons for being in Vietnam and amplifying the antiwar movement in the 1960s. Daniel C. Hallin refers to this war as the "uncensored war" because the media seemed to provide physical and concrete evidence of what was going on in Vietnam during this time. The media during this time also provided a kind of ambiguity about Vietnam that allowed the American people to both see what was going on and think for themselves. People will be affected by the news however much they allow themselves to be affected by it, and their pre-existing notions of what they know will only be emphasized by what they are seeing on television. Media did not simply report on the Vietnam War; it generated new levels of mistrust between the US government and the people.

The Vietnam War being televised caused a rise in anger from the American public, causing continued distrust between the American people and the government. From the first television news coverage of the religious prosecution of Buddhists in South Vietnam in 1963, the American people started consuming antiwar propaganda and media that circulated around the world, producing waves of protests that showed the power of televising the events in Vietnam. Due to coverage of events like the Tet Offensive, Americans felt a wave of discontent with their government and questioned their trust in policymakers and many began to place it within a small box in their living room. Many Americans lost faith in the government, subsequently producing protests and media forms that were created and stemmed from distrust for the government. Public opinion on the Vietnam War thus plummeted because of the detrimental effects that the TV had on the American living room and the families that watched them. Because the Vietnam War was televised and pushed onto the American public in a way that was easy to understand and easy to grasp, unlike wars before it, Americans questioned the costs and consequences of losing American lives in support of South Vietnam. Many no longer saw the value of American soldiers dying for political gains in the fight against communism.

As new technology develops, people want to see where it can go, and as television has developed, people have begun to be interested in not just fiction like in film or TV, but in real, human events and actions that take place on a day-to-day basis. The Vietnam War was a perfect setting for people to see what occurred on a day-to-day basis in a real-world horrific setting that existed because of real-world problems. People wanted to see what was happening in Vietnam. They did not know what the repercussions might be of seeing what they did not know at the time to be horrific and tragic events. What they found was a shattering of the glass veil between the American public and the American government. The truth became clear to American citizens, and they wanted more truth and more understanding of the situation; to which the government did not comply. I argue that the Vietnam War and the filming and broadcasting of it both decreased public opinion and increased a sense of self among Americans in the 1960s to 1970s. This showed the power that television held and the grip that it had on the American public through its means of mass communication and togetherness, which then created formed an interesting bond between two entities with a seemingly common goal.

Notes

1. William M. Hammond, "The Tet Offensive and the News Media: Some Thoughts on the Effects of News Reporting," *Army History,* no. 70 (2009): 7. http://www.jstor.org/stable/26373973.
2. Richard R.Lau, Thad A. Brown, and David O. Sears, "Self-Interest and Civilians' Attitudes Toward the Vietnam War," *The Public Opinion Quarterly* 42, no. 4 (1978): 464. http://www.jstor.org/stable/2748371.
3. Chester J. Pach, "TV's 1968: War, Politics, and Violence on the Network Evening News," *South Central Review*, vol. 16/17 (1999): 29. *JSTOR,* accessed 12 Feb. 2023. https://doi.org/10.2307/3190074.
4. Michael P. Sullivan, *The Vietnam War A Study in the Making of American Policy,* (Lexington: The University Press of Kentucky, 1973).
5. Daniel C. Hallin, *The "Uncensored War": The Media and Vietnam* (Oxford: Oxford University Press, 1986), 126.
6. Nicole Mallia et al., "Press at War," *US History Scene*, April 1, 2019. https://ushistoryscene.com/article/press-at-war/.
7. Hallin, *The "Uncensored War,"* 3.
8. Hallin, *The "Uncensored War,"* 3.
9. Hallin, *The "Uncensored War,"* 131.
10. Michael Mandelbaum, "Vietnam: The Television War," *Daedalus*, vol. 111, no. 4 (1982): 157.
11. Mandelbaum, "Vietnam: The Television War," 161.
12. Hammond, "The Tet Offensive," 8.
13. Michael C. Mitchell, "Television and the Vietnam War," *Naval War College Review* 37, no. 3 (1984): 42.

Atomic Television in the 1950s

CHRISTOPHER PORTER

In 1953 President Eisenhower announced to the United Nations, "(t)he United States knows that if the fearful trend of atomic military build-up can be reversed, this greatest of destructive forces can be developed into a great boon, for the benefit of all mankind."[1] Eisenhower's speech expressed his hopes for peaceful nuclear developments and came only seven years and four months after the United States had dropped the first atomic bomb over Hiroshima, Japan. In 1945 the United States had held a monopoly on nuclear weapons, but American scientists and engineers had already begun work on an improved version of the atomic bomb predicting that "monopoly would be short-lived."[2] In 1949 these predictions proved accurate when American officials detected an atomic detonation from the Soviet Union, marking the first nuclear test by the Soviets and the official end to the U.S. monopoly on atomic weapons.[3] What ensued was an intense nuclear stockpiling by both nations and the continued development and improvement of atomic weaponry. With the ever-increasing atomic competition between the U.S. and the Soviet Union and the successful development of a hydrogen bomb by the Soviets in August 1953, President Eisenhower's call for a reversal in the direction of atomic military build-up that same year can be understood as an attempt to position America as a productive force for peaceful atomic development while portraying the Soviets as its potential abuser.[4]

Eisenhower's 1953 speech was not the first instance of peaceful uses for atomic power being proposed. In fact, such ideas were prevalent ever since the discovery of nuclear fission in 1938 when the first hopes were to harness the released power for energy, not destruction.[5] The escalation of the wars in Europe and the Pacific in the early 1940s shifted research in the direction of a new weapon. After the United States dropped atomic bombs on Japan, the dichotomy of the atom as a weapon capable of leveling a city and at the same time a possible source for almost ceaseless power became central to American debates about the future of atomic power in domestic politics and on television. Between 1950 and 1959 televised debates and political broadcasts, as well as entertainment and informational programs, repeatedly brought the topic of nuclear weapons and energy directly to the American public. Indeed, this paper will show that broadcasts became a tool for framing the American perception of the atom throughout the 1950s.

This research is interested in how television in the United States reflected and reinforced both American fears and optimism surrounding nuclear capabilities in the 1950s. This question has not been extensively explored, and it brings together two separate literatures: historical research on atomic energy and media studies of television in the 1950s. Scholars that have analyzed postwar atomic research have looked at the different applications of radiation and atomic concepts in science for both civilian and governmental purposes. Several of these analyses focused on Cold War studies of radiation as a cure for cancer and a tool for improving agriculture, as well as government projects focused on controlled nuclear detonations for the purposes of altering landscapes to build canals or collect fossil fuels.[6] Such analyses include Jacob Hamlin's *The Wretched Atom,* Helen Curry's *Evolution Made to Order*, and Angela Creager's *Life Atomic,* each of which are referenced within this work and are essential to better understanding the American attempts at creating new uses for the atom throughout the 1950s.

By contrast, media studies research on television in the 1950s has focused on analyzing elements of popular television shows, how America's growing consumer culture was represented on television, and examining the "oligopolistic control" of the three major broadcasting networks.[7] This paper is most interested in the research that has been conducted on popular elements of television shows such as those discussed in Cynthia Miller and A. Bowdoin Van Riper's, *"Rocketman" TV Series and Their Fans: Cadets, Rangers, and Junior Space Men,* as well as Rodney Hill's *The Essential Science Fiction Television Reader*. Each of these resources factor into my own analysis of science fiction television in the 1950s as these scholars provide helpful contextual understanding of specific shows like *Space Patrol* and *The Twilight Zone* through discussions of their individual messages and analyses of the followings they had.[8]

Where most scholars have focused on the merits of individual uses for atomic research or individual television series in the 1950s, my focus combines these topics and analyzes how the atom was discussed on television throughout the 1950s. I argue that throughout the 1950s, when discussed on television, atomic concepts were framed in such a way to only permit one view: that America does and will use the atom in productive ways while its use by other nations should be viewed as a threat to Americans. This was largely accomplished by framing the atom to only be understood as either a source for nearly endless power, or as a threat to American safety when wielded by enemy nations. This framework appeared in political and entertainment television throughout the decade, each of which explored discussions and representations of radiation, nuclear bombs, and atomic research as concepts that could either be controlled by man for productive uses or abused in largely destructive ways. Television programming consistently presented Americans as productive users of abundant energy or the unwitting targets of an attack. This dichotomy, coupled with the undermining of individuals who challenged the pro-American narrative, was a significant factor in making television a closed forum for discussions of atomic concepts.

This essay has been organized into two sections: "Politics on Screen" and "Entertainment Television." This ordering is not a chronological reflection of the atomic debate; rather my focus is on highlighting two different forms of television, each with their own distinct viewership, to demonstrate that the discussion of the atom and America's relationship with it was consistent across different types of programs throughout the decade. This chapter begins with political television to introduce and discuss the political climate and programs of 1950s America surrounding the atom. While some of the entertainment television shows appear at the start of the decade, they are featured second in the paper for conceptual cohesion to keep the focus on how the pro American atomic message appeared in each distinct type of broadcast, rather than emphasizing a strict chronological order.

POLITICS ON SCREEN

Between 1950 and 1959 scholarly and political discussions surrounding nuclear power were semi-frequent on American television sets. During this period high profile incidents fueled popular interest in the topic, from President Truman's televised address announcing that the U.S. government had detected a successful Soviet nuclear test in 1949, to Eisenhower's "Atoms for Peace" speech in 1953, and the successful flight of Sputnik (a Soviet satellite sent into low earth orbit that was capable of being equipped with a nuclear warhead) in 1957. Such broadcasts kept American awareness surrounding nuclear technology peaked, but they also drove an ever-increasing fear of nuclear destruction. Each new advancement in Soviet and American technology, from the successful Soviet detonation in 1949, to U.S. nuclear tests in Yucca Flats (see Jeremiah Albert's contribution in this volume), and the development of hydrogen bombs in the early 1950s, intensified the attention to and discussion of nuclear power and how to live with it.

Contemporary political shows, including televised political speeches, debates, and news broadcasts, comprise an archive of the evolution of American goals for nuclear power. The Soviet breakthrough in 1949 was just the first of Soviet advancements that motivated the U.S. government to make concerted efforts towards advancements in peaceful uses of the atom. However, advancements and discussions of peaceful uses for atomic energy did not mean the American government moved away from atomic weaponization; rather Soviet advancements also served as a motivating factor in America for the further stockpiling of nuclear weapons. The American message of peace and the Soviet nuclear program therefore mutually served American interests by allowing the U.S. government to be recognized internationally as one voice behind a movement for progressive demilitarization while also appearing justified in their growing weaponization of the atom for claimed defensive purposes. Several of the peaceful projects the U.S. pursued included interests in sustainable nuclear power for cities, radiation treatments and research to advance medicine and agriculture, and controlled nuclear detonations to expose and obtain natural gas.[9] Many of these topics became talking points for political speeches and broadcasts that emphasized the ability to repurpose the atom from a weapon to a tool for scientific advancement.

Most of these efforts were discussed to some extent on American political television like *Meet the Press*, a political broadcast which brought together representatives from multiple news agencies to discuss pertinent issues with someone involved in the field being referenced. However, no amount of representation of peaceful uses for the atom could eliminate the similarly prevalent image of the destructive atom, constantly referenced in the same political broadcasts as well as safety training videos for school children.[10] This ensured American viewers were constantly receiving a split perception of the atom as a tool for peace *and* as a weapon of mass destruction. Discussions of the destructive capabilities of the atom often referred to the threat of the Soviets holding nuclear capabilities or the strength of American stockpiles to deter attacks. Thus, a secondary message in American political television emerged of America as peacekeeper and seeker, and the Soviet Union as a threatening foe within the context of the Cold War.

America in the 1950s showed an active interest in peaceful uses for atomic power, but the atom's destructive capabilities remained a prominent focus of all atomic conversations. The same speeches and shows which discussed hopes for harnessing nuclear energy in a peaceful manner contrasted with speculations about military advancements and concerns over Russian nuclear capabilities. This duality was a major factor in ensuring that the American public saw the potential for *American* pacification of the atom, while similarly recognizing the potential for other countries to wield it against them. The destructive potential of the atom, therefore, was advanced in political television either to promote fear of other nations developing nuclear weapons or reassure American audiences of the *defensive* strength of America's nuclear arsenal.

President Eisenhower's *Atoms for Peace* speech in 1953 presented possibly the prime example of the dualistic political message. While the primary emphasis of Eisenhower's speech was on hopes for peace and cooperation between all nations, he did not miss the opportunity to assure his audience of America's nuclear strength. He reminded Americans that "The United States' stockpile of atomic weapons, which, of course, increases daily, exceeds by many times the total equivalent . . . of all bombs and all shells that came from every plane and every gun . . . in all the years of the Second World War."[11] Eisenhower went on to say that atomic weapons "have virtually achieved conventional status within our armed services" and that every branch of the United States military was "capable of putting this weapon to military use."[12] While it seems to be a strange prelude to a message of hope and peace, Eisenhower clearly attempted to represent American strength and to deter other nations from mistaking his call for peace as a sign of American susceptibility to a nuclear strike. Indeed, Eisenhower's speech communicated a split message to two different audiences—internationally and to the American public—presenting the U.S. government as the arbiter of progressive demilitarization in favor of peaceful nuclear advancements, while similarly reassuring that same audience of America's nuclear strength that could be utilized in defense if necessary.

Reporters and newscasters had a similar impact in shaping the perception of America's relationship with the atom by leading authoritative discussions on nuclear energy. Many of these prominent voices controlled the narrative that was presented to their audience in how they framed their questions and responded to guests. One such broadcast aired on May 11, 1958, in an interview with Dr. Linus Pauling, an American chemist who had been awarded the Nobel Prize for chemistry in 1954, on an episode of *Meet the Press*.[13] Dr. Pauling had been brought on the show as a representative of a growing movement in the U.S. to bring an end to nuclear testing and as a respected scientific researcher with experience in the effects of radiation on humans.

The Pauling interview represents how political television was often mobilized to communicate and defend a specific view. The interviewer who introduced Dr. Pauling frequently directed Pauling into making claims he seemed hesitant to assert or brought his expertise into question based on outside factors. At one point in the interview, for example, Pauling attempted to convince the audience that conventional statistics comparing radiation from nuclear tests and natural radiation were misleading and undermined a realistic understanding of the harmful effects of nuclear testing. In response, the interviewer asked Pauling if he believed the audience would trust his opinion if they were also aware of his connections to the communist party. This redirection is monumental: the interviewer began to challenge Pauling's expertise compared to others in the field and even used alleged communist affiliations to undermine his warning to the audience.

Similarly, Dr. Pauling began the interview with his call to end nuclear testing but was quickly redirected into talking about the strength and importance of nuclear capabilities by one of his interviewers, an American publisher named Lawrence Spivak.[14] When asked if he thought that "the security of the free world is dependent upon the deterrent power of the United States," Dr. Pauling hedged that "the security of the world is dependent now upon the deterrent power of nuclear weapons, and we have the greatest stockpile . . . in existence."[15] Pauling was clearly measuring his words in response to the question to avoid offering a defense for America's own nuclear stockpile. Rather, his response acknowledged that in the state of the world as it was then, with the rapid production of nuclear weapons from the Americans and the Soviets, deterrence played a factor in preventing war and a nuclear strike, but it also heightened the potential for an accidental attack.[16] Pauling felt it would be best if, over time, nuclear arms could be diminished or abandoned wholesale. However, Spivak's attempts to force Pauling into answering restrictive questions, such as if America's stockpile was an effective deterrent or if radiation from nuclear testing was more dangerous than natural radiation, caused Pauling's own optimism about the atom to be obscured. Instead, the message that rang out most clearly to the audience was that America's nuclear stockpile is an effective deterrent against nuclear warfare, and that there was debate over how dangerous radiation from nuclear testing truly was.

Political speeches and talk shows were not the only realms in which the perception of the destructive atom pervaded television; it also became prevalent in schools with the introduction of "Duck and Cover drills" in 1951. The "duck and cover" campaign, initiated by the Federal Civil Defense Administration, sought to prepare American school children for the possibility of a nuclear strike. The campaign involved teachers showing children a television ad focusing on a turtle named Bert, who showed students how to duck and cover to defend from a nuclear explosion. Bert, of course, carried his own shelter around with him in the form of his shell and was backed up by a narrator who informs students that when a nuclear bomb goes off, "if you are not ready and did not know what to do it could hurt you in different ways. It could knock you down hard or throw you against a tree or a wall . . . it can smash in buildings . . . but if you duck and cover like Bert you will be much safer."[17] This ad acknowledged the atom's destructive capabilities, if not its often lethal nature, and it made clear that the U.S. government was concerned enough with the possibility of a nuclear attack from another nation that they initiated school drills to prepare for such a strike. This ad campaign made it impossible to look only at the peaceful capabilities of the atom when any day at school could involve preparing for a nuclear attack.

Political shows and public service campaigns such as Duck and Cover discussed the atom and the state of nuclear developments with an air of authority that audiences could trust. Through such authoritative messages, however, the atomic debate was often simplified, constructed to reflect America as promoting peace while stockpiling weapons to prevent an attack from other nations. In the examples above, Eisenhower's mention of the growing nuclear stockpile served as a message of deterrence amidst a call for peace, while Pauling's beliefs were overshadowed by an interviewer intent on defending American nuclear stockpiling and testing, and "duck and cover" served as a direct representation of the American government striving to inform its people on how to protect themselves in the event of an attack. The "destructive atom" was condemned as a tool of opposing nations when discussed in 1950s political television, yet simultaneously it was hailed as a great defense in the hands of Americans who would use it wisely. Many of these broadcasts split the discussion of the atom between its peaceful or its destructive applications, but the nuclear intentions of Americans always were represented or defended as being for protection and scientific advancement.

The Peaceful Atom

The atomic bombing of Japan demonstrated in dramatic fashion the destructive potential of nuclear energy, and American leaders interested in potential civilian uses of the technology worked to contain public opinion and push forward with research and development. Almost immediately after the Hiroshima and Nagasaki blasts in August 1945, American officials renewed calls for peaceful applications of nuclear energy. When President Truman first told the American public about the destruction of Hiroshima, he made sure to also mention the potential that came from releasing the power of the atom as a possible alternative to "coal, oil, and falling water."[18] Soon thereafter Truman established the Atomic Energy Commission to "manage the development, use, and control of atomic (nuclear) energy."[19] Truman's convictions about the potential peaceful uses of the atom became similarly prevalent in the 1950s and the subsequent administration of President Eisenhower.

President Eisenhower's "Atoms for Peace" speech served as an advertisement for the benefits that could come from further developing peaceful uses for the atom. Eisenhower framed the discussion of the peaceful atom in the international community as a means to begin reducing nuclear weaponization while also bringing new technology and sources of energy to developing nations. He called for the formation of an international atomic energy agency that would collect donations of uranium from nations with existing stockpiles of the resource to be protected and studied by the members of the agency for the purpose of serving "the peaceful pursuits of mankind."[20] Eisenhower's speech argued that if nations could band together "in good faith" to more rapidly develop atomic energy, that resource could be shared with developing countries and progress could be made toward beginning "to diminish the potential destructive power of the world's atomic stockpiles."[21] Eisenhower's speech was a clear call to strip the atom of its destructive imagery and instead embrace wholeheartedly its power as a peaceful tool for the advancement of humanity. At the same time, Eisenhower's speech effectively framed America as a leading voice in the call for nuclear reduction, an important development that began to show an American desire to distance itself from its nuclear strike on Japan and force Soviets to either echo or reject the American call for nuclear peace.

Other American politicians and political talk shows repeatedly echoed Eisenhower's peaceful message following his speech. In fact, only one day later Representative Carl Hinshaw (R-California) appeared on the show *Chronoscope* (CBS) with famed newsman Larry LeSueur to voice support for the president's speech.[22] Representative Hinshaw expressed his own optimism about the willingness of the U.S. to share its nuclear secrets with other countries that agreed to do the same.[23] His optimism extended even to the Soviets who he hoped "have some information which would be just as valuable to us as our information would to them."[24] This message alone was an important note as once again an American politician is seen advocating for openness and collaboration as it pertains to atomic developments and calling explicitly on the Soviet Union to do the same. If the Soviets agreed to the sharing of information, the United States would have succeeded in its call for collaboration; if the Soviets refused, the Americans again would come out on top as the promoters of a peace rejected by the Soviets. This public call for peace framed America as willing to come to the table and move away from weaponization while putting all the pressure on the Soviets to do the same. If the Soviets would not, then that fact alone becomes a justification for America's continued stockpiling, maintaining the image of America as the benevolent user of the atom.

In his interview, Hinshaw was most effusive about the peaceful potential for nuclear energy: he believed the U.S. was very near harnessing this energy, and he claimed it would be a great benefit to countries such as India that "are suffering from a shortage of power."[25] Hinshaw perfectly articulated the crux of the debate when he claimed "fissionable material may be used for an explosion or it may be used in a what you might say a slow explosion." Hinshaw's "slow explosion" involved setting off an atomic reaction in a chamber that would allow scientists to control the rate at which the particles collided, effectively prolonging the explosion, allowing scientists to harness the energy produced.[26] Hinshaw claimed that if scientists could focus wholly on the slow explosion for the production of energy, the industrialized world could repurpose all nuclear weapons toward peaceful uses instead.[27] While Hinshaw speaks about his hopes for collaborative international efforts, his position as an American politician presenting hopes for the peaceful production of nuclear technology further serves the framing of America as a positive influence on atomic developments rather than a perpetrator of its weaponization.

Milestones in the potential development of nuclear weapons continued to be met with counterexamples of the peaceful potential of nuclear energy in the American press. In 1958, just one year after the successful launch of the Soviet satellite Sputnik, conceptions of atomic peace permeated the press once again. This time, government-adjacent spokespeople comprised the dominant voices. Dr. Pauling's 1958 interview serves as a prime example of the peaceful atom's appearance and framing on television. Pauling's call for an end to nuclear weapons testing, echoed the sentiments of a growing movement against nuclear weapons in the United States.[28] He noted transnational interest in nuclear disarmament, citing both American and Soviet proposals to do so in 1957 and 1958 respectively.[29] Dr. Pauling's hopes for disarmament, which involved seeking a resolution to international conflict without war to avoid the potential for accidental nuclear strikes, were reflective of a growing sentiment among U.S. citizens and showed the extent to which the language and principles of the peaceful atom had spread since Eisenhower's "Atoms for Peace" speech just five years earlier.[30]

This is only a small sampling of the vast number of interviews televised during the 1950s concerning the atomic question. Across the decade, arguments that the atom was capable of peaceful and productive advancements were actively reinforced by politicians and experts, who suggested that once the science for peaceful uses of the atom was settled, the plutonium in existing weapons could be repurposed. Television interviews with government officials and their spokesmen argued that the U.S. government had been engaged in pursuing peaceful uses of the atom since the late 1940s with varying success.[31] These arguments were mostly accurate as the federal government had in fact dedicated multiple resources to the exploration of non-weaponized atomic developments with projects that carried on well into the latter half of the century, although many proved fruitless.

Research into atomic applications spanned several different aspects of society ranging from to medicine to energy and agriculture. Developments in medicine included the federal government's active sponsorship of "biomedical research" that involved radiation, such as looking into its possible use as a cure for cancer.[32] Despite the early conclusion that radioisotopes were "causes rather than a cures of cancer," the government continued backing further research into the medical uses of radiation and found that it could be "a valuable tool in investigating disease."[33] Other avenues of atomic research involved the irradiation of seeds with the goal of producing useful mutations in plants, the development of nuclear power plants for the harvesting of nuclear energy, and tests of controlled detonations for shaping landmasses to create canals, or expose and collect fossil fuels.[34] The U.S. government showed through these projects that it was dedicated to identifying positive uses for radiation and nuclear technology outside of its weaponization.

Most of these efforts, however, proved fruitless and even dangerous. In particular, little was known about the health risks of radiation in the early 1950s. The Atomic Energy Commission "and its advisors believed the benefits of atomic energy outweighed the costs in terms of health risks or environmental contamination."[35]This was an optimistic conclusion for the AEC, which continued to embark on and sponsor a great deal of scientific research into alternative uses of fissionable material. While proposals for a peaceful atom such as these were undoubtedly in pursuit of noble goals—cheap, abundant energy, breakthroughs in medical treatments, and new methods of maximizing agricultural output—it is important to remember the U.S. government never actually abandoned the destructive capabilities of the atom and did not shift away from the continued advancement of its nuclear stockpile.[36] What mattered was that publicly the U.S. was promoting a vision of a peaceful atomic world, a message that was actively reinforced to U.S. citizens through political television. Meanwhile, the American nuclear program grew and was explained away as necessary for the defense of the nation.

Programmers demarcated political television as a forum of truthfulness and public debate with an air of authority imbued in part by the experts and officials that appeared to make their cases. By contrast, entertainment television, such as science fiction and anthology shows, were vehicles for attracting large audiences to appealing programming in the service of sponsors and advertisers. These shows had a broader audience, appealing to children, teenagers, and adults, while also filling prime airtime in the evening when children were home from school and parents were back from work.[37] Made up primarily of fictionalized narratives, centered occasionally on socially relevant issues, entertainment-based shows may not have shared the authority of political broadcasts, but they offered something different in their ability to visually construct scenarios and events on which news stations could only speculate. While politicians presented authoritative statements about the atom, entertainment television dramatized its power and the implications of both its destructive and peaceful uses in a manner more easily understood by the average viewer. As viewers became habituated to welcoming entertainment television into their lives, it began to shape their perceptions and understanding of many socially relevant topics despite the medium's fictionalized nature.

The events of the 1950s as described above similarly drew the attention of producers of entertainment television to the topic of atomic energy, peaceful and destructive. Science fiction and anthology shows like *Space Patrol* (1950-1955), *Tales of Tomorrow* (1951-1953), and *The Twilight Zone* (1959-1964), presented the atom as a destructive force that could be used against Americans in devastating ways. Educational programs such as *Watch Mr. Wizard* portrayed the current science behind the atom, explaining its dangerous capability, but also its potential usefulness when put to peaceful applications. Walt Disney's *Our Friend the Atom,* one in a series of Disney weekly specials, explored the history of atomic developments and speculated optimistically about how nuclear energy could move humanity forward in science and medicine. These shows mirrored the split discussion of the atom that was present on political television, each presenting either the stance that the atom was a tool for peace, a weapon to be feared in the hands of other nations, or sometimes both.

These shows appealed to children as well as adults, both men and women, and through their use of storytelling they depicted the capabilities of the atom, both good and bad. More research needs to be done to determine how, if at all, these shows may have swayed public opinion on how the atom should be used. But in the larger political context of the US policy goal of abundant energy, and intensification of the Cold War, entertainment television confronted its massive viewership with few possible interpretations of the nuclear debate. These interpretations consistently reinforced the same narratives present among political broadcasts of the decade, depicting America as a nation devoted to peace yet susceptible to attack if other nations continued to develop their own nuclear programs.

Anthology and Science Fiction

Science fiction and anthology shows had been prevalent on television for years yet largely had been recognized as children's shows. This perception began to change in the 1950s when shows like *Space Patrol* (1950) *Tales of Tomorrow* (1951) and *The Twilight Zone* (1959) began to grip adult audiences as well.[38] These shows represented some of the most popular anthology and science fiction series of the 1950s, and each devoted at least one episode to the atom or atomic concepts. All three series primarily focused on depicting the atom as a threat which the American public should fear in the hands of countries that would seek to wield it against the United States.

Space Patrol is one of the best examples, not only of the atom becoming a prevalent topic in entertainment television, but also of the success of science fiction as both a children's and adult genre. A live action series broadcast live 1950–1955, *Space Patrol* was regarded as one of the most popular science fiction shows of the decade. The show followed Commander Corey and Cadet Happy as they traveled through space battling crime.[39] Creator Mike Moser sold the show concept to KECA, a local "Los Angeles affiliate" of

ABC (now holding the call sign KABC-TV), which broadcast *Space Patrol* as a daily 15-minute special.[40] Nine months after its debut, it had become so popular that ABC transitioned it to a 30-minute "weekly broadcast" nationwide.[41] They marketed the show largely toward children, and, in keeping with the sponsorship model of the day, many episodes featured an ad placement for Nesquik products, a favorite children's drink company. The show reached far beyond its intended audience, and viewership data indicated that the show became a favorite across a large span of age groups.[42]

In 1954, just two months after Eisenhower's Atoms for Peace speech, the show took up the issue of nuclear energy. In a two-part episode—"The Deadly Radiation Chamber" and "The Plot in the Atomic Plant"—Commander Corey and Cadet Happy find themselves on a planet with cities powered entirely by nuclear energy.[43] They face an unknown foe revealed to the audience to be Professor Proteus, a trusted ally of the two heroes and a regular on the show, who wants to control the reactor and scare off any competition. To do so, he spreads rumors that the planet's nuclear reactors are failing and emitting mass amounts of radiation.[44]

Commander Corey and Cadet Happy investigate these claims and become trapped in a radiation chamber with which Professor Proteus, disguised as a janitor for the power plant, attempts to kill off the heroes with radiation poisoning.[45] Slowly, the two heroes begin growing visibly tired and weak until Commander Corey realizes he can pull out the reactor's control rods, opening a slot in the wall through which they can escape.[46] They are soon discovered by their sidekick Carol Carlisle and taken to a hospital for the "cure."[47] The process of curing radiation poisoning is not explained or depicted; it is simply stated that this society has the ability to cure radiation as long as those affected are treated quickly enough. Following these episodes, Professor Proteus continues trying to thwart Corey and Happy in his many disguises while the heroes attempt to find out who is behind the radiation rumors.

These two episodes posit a positive hope for advancements in nuclear energy and medicine as the result of further exploring the possibilities of the atom. What American audiences see from this episode is a possible future in which atomic energy powers their cities and any harmful radiation that is produced could be cured. The atom is depicted as controllable by man with the only dangers coming at the hands of nefarious actors, not the unpredictability of the resource itself. When an outsider like Professor Proteus seeks to use the atom for personal gain, it can be used as a deadly weapon against the forces of peace, like our heroes Commander Corey and Cadet Happy, who are clearly coded as American. This is the precise mirror of the narrative that so frequently appeared later in political broadcasts during the 1950s that Americans represented a force for peace and could be trusted with nuclear technology, but outsiders would attempt to abuse it in harmful ways if left unchecked.

However, what was possibly the most frightening aspect of these episodes for the viewing audience was that the radiation was invisible. The show did present a future in which all radiation can be cured if treated quickly enough, diminishing how threatening harmful radiation may appear to an uninformed audience, but its invisibility still presented reason for concern. Corey and Happy were aware of the danger because they saw the rods and understood what was happening, but the audience could be left to wonder how they would ever know if they had been exposed to deadly amounts of radiation in a world in which nuclear energy was being increasingly pondered.

These two episodes dramatized for a concerned audience the possibilities of a nuclear future. On the one hand, the show depicts the dream the American government seems to promise with nuclear energy and vast advancements in medical treatments, but on the other, it depicts a world powered by a force that could be manipulated if left in the wrong hands. This split image helps to further frame for the American audience that the goals of the U.S. government to develop peaceful atomic technologies is one worth investing in so long as the U.S. remains strong enough to keep others from abusing such technologies, especially against America. That strength typically coming from America's own nuclear arsenal.

A second show important to the atom's depiction in entertainment television was *The Tales of Tomorrow* (1951-1953, ABC), a widely popular anthology series. Standalone episodes focused on classic stories like Frankenstein, as well as speculative accounts of future events. In its first season, the show played off the rising wave of anti-communism in the United States as well as contemporary nuclear fears in an episode that took its inspiration from the Japanese attack on the United States. "Sneak Attack," broadcast on the tenth anniversary of Pearl Harbor, centered on an unfolding crisis in which several unmanned mystery planes have landed in American cities.[48] When one of the planes explodes, its nuclear payload wipes Denver off the map, killing 46,000 and wounding more.[49] Shortly after the explosion, it is revealed by the Soviets, who had earlier that same day discussed peace with the President, that they are responsible for the attack; if the Americans do not surrender within four hours, they assert, the others will explode as well.[50] Ray Clinton, an American spy in the USSR on a reconnaissance mission, who has been detained and surveilled in a Soviet hospital, becomes the last hope for preventing the destruction or submission of America. Ray learns through one of the nurses in the hospital that the Soviets have threatened America and that there is only one hour left to prevent any further explosions.[51] Ray manages to contact the American Chief of Staff instructing him not to surrender because he believes he can shut down the Soviet planes from within the hospital.[52] With the aid of the nurse, Ray neutralizes the guards, shuts off the bombs, and saves the day for the Americans.[53]

Though the scenario in "Sneak Attack" is clearly fictionalized (and highly improbable) if not for the plot then for the resolution, the conflict of the episode reflected a very real American fear that the Soviets would reach a breaking point during the Cold War and launch a nuclear attack. What made this episode particularly frightening was that, without Ray (representing American infiltration of the Soviet Union), America had no hope of retaliation or defense; the doctrine of deterrence had failed. Indeed, in the episode the secretary of defense informs the president if the other bombs were to detonate, the United States would suffer "fifteen million dead, eighty percent of our industry shattered, and sixty-five percent of our land leveled," with no hope of launching a counter strike.[54] The president frames the question of surrender as a decision between "doom" and "slavery."[55] This episode is a complete break from the hopeful and peaceful representations of the atom that had been promoted in other shows and political broadcasts. Instead this was a frightening glimpse into a bleak future for the United States if the Soviets or other enemy nations were left unchecked in their development of nuclear arms.

The airing of "Sneak Attack" on the ten-year anniversary of Pearl Harbor only further heightened fears that a scenario like the one depicted in the episode could come to light during the Cold War, evoking a sense that if it happened once it could happen again. With the Soviets having achieved nuclear capability two years prior and the tensions of the Cold War constantly increasing, there was certainly a widespread fear that a nuclear strike may come without warning against the U.S. The episode itself hinges on deceit as the Soviet ambassador had that same day engaged in peace talks with the American president before the Soviets threatened to level America if its government did not surrender. The U.S. is cast as the defenseless target, much as it was during the attack on Pearl Harbor. However, unlike Pearl Harbor, this situation is resolved with only minor Soviet casualties due to the quick actions of American spy Ray Clinton. The show, then, represents the United States as peace-keepers and -seekers while it represents the Soviets as deceptive and aggressive in their use of atomic technology.

Lastly, at the end of the 1950s, CBS premiered the *Twilight Zone*. The show quickly became a hit and ran for five seasons from 1959 to 1964.[56] *The Twilight Zone* was another anthology series that presented realistic but also many otherworldly alternate realities to the audience. In its first season, just one year after Sputnik, the show aired "Time Enough at Last." This episode centers around a meek bank teller named Henry Bemis who wants nothing more than some quiet time to read but is constantly kept from his books by his demanding manager and his overbearing wife.[57] After being threatened with the loss of his job if caught reading again, Henry sneaks into the bank's vault during his lunch break to read the newspaper without any prying eyes.[58] Notably, he seems to overlook the headline "H-Bomb Capable of Total Destruction."[59] Only moments later he is frightened by a violent earthquake that shakes the vault.[60] He cautiously leaves his hiding place to discover that the world around him has been devastated: an H-Bomb has struck his city leveling buildings and obliterating all life.[61] Only Henry survived due to the protection of the vault. Henry explores the wreckage, finding a library with enough books and peace to read for the rest of his days just before the episode ends on an ironic note when he drops and breaks his glasses.[62]

While the conclusion of the episode is comical, the complete destruction of the world outside the vault suggested how devastating a single bomb could be. Here again Americans were presented with an image of the world they knew, a city that stood in for any around the country, and in seconds everything was destroyed by a bomb dropped by an unnamed adversary. This episode, perhaps more adequately than those in *The Tales of Tomorrow*, reflects American fears that a nuclear strike could happen to the U.S., initiated not just by the Soviets but from any nation with nuclear arms. Here again was a representation of the atom as a destructive threat that could be used against America by any nation who would not wield it in the peaceful manner Americans advised.

Space Patrol, *Tales of Tomorrow*, and *The Twilight Zone* were not the only entertainment shows from the 1950s that reinforced the dual image of the atom or even featured nuclear concepts within their episodes, but they do represent three of the most popular examples. With high weekly viewership it was certain that these shows confronted audiences across the nation with the dilemma of the nuclear world. These shows projected the possibilities of the nuclear future, suggesting what the world could look like if the hopes or fears expressed by politicians about nuclear power were realized.

Storylines centered on the fear of a nuclear attack on the U.S. or the hope for peaceful progress through the development of nuclear energy; as such, American audiences were given only two perspectives from which to view the atom. Further, entertainment television consistently posited a world in which Americans were the victim or potential victim of a nuclear attack. So, not only was the idea that the atom could only be used for scientific advancement or destruction present in entertainment television, but the concept of Americans as protectors of the peace while outsiders, especially the Soviet Union, were threatening foes to Americans was heavily prevalent as well.

SCIENCE ON AIR

Another forum for the discussion of atomic energy on television that captured a large audience during the 1950s was educational programming such as the children's science show *Watch Mr. Wizard* or the hour-long Walt Disney special, *Our Friend the Atom.* These shows presented scientific and policy information in an entertaining manner that was lacking from news or political discussion shows. They appealed largely to younger audiences but provided something even parents could enjoy through the educational nature of the program itself. These two shows are significant because they each essentially bridged the gap between the more narrowly political and entertaining (science fiction) presentations of the atom. These shows had an air of authority that was missing from science fiction shows but maintained a level of excitement that would bring in wider audiences uninterested in politics. By capitalizing on the most appealing and influential aspects of the other two genres, educational television added another strong layer of influence to the framing of the atom on American television.

Science television gained a growing audience in the 1950s. In 1957 NBC released a program airing on Saturday-mornings called *Watch Mr. Wizard,* which "aimed to inspire viewers to become amateur scientists and to explore science in their own life."[63] The show was based on the activities of a young scientist, who welcomed the neighborhood kids into his home to help him with experiments, while he taught them about the science behind what they were doing. This made science fun, interesting, and accessible to young audiences, who could occasionally replicate his experiments at home and pass on what they learned to their parents. Despite the invocation of the wizard figure, the show "carefully delineated the difference between magic and science," ensuring its audience understood that what they were being taught was always factual.[64] Mr. Wizard became someone kids could trust, a fact that ensured children would find him credible when he spoke about the atom and its potential.

In 1953, NBC aired an episode that focused primarily on the atom. In "Atomic Energy," Mr. Wizard explains nuclear concepts to one of the neighborhood boys, Willie, by defining such terms as "atom" and "electron," and representing the concepts visually. The show opened with Mr. Wizard explaining a chain reaction by lighting a branching line of gunpowder and watching as the flame spread down the path flaring up at dividing points.[65] Mr. Wizard uses the chain reaction as a launching point into his explanation of the benefits of atomic energy. Having demonstrated the release of energy via the chain reaction, Mr. Wizard switched gears to focus on where that energy might be found. He presented Willie with a pound of butter saying if we could "change that to electric current and could sell that current at the rate that you're paying for electric energy today," it would be worth about 110 million dollars and "could run all the railroad trains in the United States for a couple of years."[66] Mr. Wizard presents to the audience in a simplified way the power and value of atomic energy, elevating its allure as a product worth further exploring.

Interestingly, Mr. Wizard briefly explained to Willy the concept of the nuclear bomb but did not linger much on the topic nor did he deeply discuss its devastating potential. Rather, he kept only to the scientific facts a child may need to know to understand scientific concepts that underpinned the bomb. Such facts mainly included the definition of the terms Mr. Wizard had introduced Willy to at the outset of the episode. Thus, in presenting the nuclear bomb to his audience, the Wizard minimizes the destructive potential of the atom, in favor of its potential significance as a tool of economic and industrial prosperity.

The episode reinforced the more peaceful understanding of the atom, urging its audience to view nuclear technologies as something to be further studied and harnessed for the benefit, and profit, of mankind. Not only that, but it similarly presented the notion that energy could be found in any benign item, not just plutonium. This fact amazed children, and likely adults, with the possibility that such objects – including any pound of butter – could be worth 110 million dollars in the energy industry. More generally, what Mr. Wizard presented was the idea that atomic growth in non-militarized technologies could benefit all industrialized nations. So, again, American audiences came to believe that American scientists conceive of the atom as a tool for peace and that are making the effort, or the call, for its development.

While *Watch Mr. Wizard* downplayed the nuclear bomb as but one outcome of "science" for a young audience, Walt Disney's *Our Friend the Atom* was more intentional about representing the non-military uses of the technology. This special was the most direct example of a television program creating a dual image of atomic power. Disney created the film at the request of the Eisenhower administration to depict the "peaceful applications of atomic energy."[67] Eisenhower's goal was to reinforce his "Atoms for Peace campaign" by showing viewers the many potential uses for, and benefits of, atomic energy.[68] What makes this specific program an excellent example of splitting the perception of the atom is that Disney does not only provide hope for future peaceful advancements, but his program also lingers with a tone of regret on the use of nuclear weapons in World War II and the destruction they caused – one of the few such representations from contemporary television.

Our Friend the Atom combined animation and reality to make atomic concepts more accessible to the average viewer. Like many Disney specials, the program invoked well-known stories to contextualize the science presented by "experts" speaking with an authoritative mode of address. The special opened with a discussion of *Twenty Thousand Leagues Under the Sea* and an appearance by Walt Disney himself. Disney informed the audience that fiction had become fact with the first atomic powered submarine, the nautilus, reproducing what had been imagined in Jules Verne's story.[69] Similarly, Disney's expert, a real scientist called Heinz Haber, compared the discovery of the atom to the story "Arabian Nights," about a fisherman who finds a vessel containing a genie.[70] As Haber put it, scientists were like the fisherman, uranium the vessel, and the atom was the genie with the potential either to destroy or to grant wishes.[71] So, from the very outset of the program, Disney made explicitly clear to the audience that the atom could only be understood either in terms of destruction or in terms of scientific advancements.

Staying true to this point, Haber explained atomic progress over time and more specifically the concept of an atomic chain reaction. In an animated segment, one molecule breaks loose from an atom of uranium, colliding with the nucleus of another atom of uranium and sending two more molecules off to repeat the process.[72] Haber represented the action physically before the episode transitioned into multiple clips of nuclear explosions; he stated that it was with the dropping of the atomic bomb that the genie was released and "posed a fearful threat."[73] But Haber explained to the audience that, if we could change paths away from further weaponization and focus on slowing down and prolonging atomic chain reactions, we could instead harness the atom's destructive power as energy to replace fossil fuels, power our cities, our planes, our submarines, and even utilize its radiation to make advancements in medicine and agriculture.[74] Haber claimed that atomic reactors were the key to harnessing atomic energy and that such reactors "give us a chance to make the atomic genie our friend. He will come forth to our beckoning and . . . grant us three wishes" says Haber.[75] The story Disney tells is one which blends fantasy and reality but also intrinsically establishes the atom as either destructive or productive and controllable by us.

Between *Watch Mr. Wizard* and *Our Friend the Atom*, American audiences were presented a new form of atomic representation which combined the facts of political television with the entertainment value of science fiction to produce another convincing forum to mold American perceptions of the atom. Both programs referenced the atomic bomb and shared hopes for the future of atomic energy describing it as a fantastical thing which could produce not only vast monetary profits, but also nearly boundless energy. Between them, America was the voice of hope for atomic progress that could benefit the world, but also more directly benefit major markets in American society. Together, these two programs nicely combine to further establish the dualistic image of the atom which was permeating in American television during the 1950s.

CONCLUSION

In the 1950s, television programming framed all conversations about the atom in only two ways, either as a source of nearly endless power to be actively pursued, or as a threat that the American public should fear when wielded by countries other than America. Political shows like *Meet the Press* and *Chronoscope* were two important forums for this conversation. Hosts and guests grappled with the implications of nuclear energy, considering its potential in a multi-polar world and whether Americans should be concerned about making nuclear concessions when the Soviets had a similar nuclear capability. A consensus emerged that nuclear power would be central to postwar American economic prosperity and, while the shows did entertain discussion of the destructive potential of the atom, they also contained possible anti-nuclear interpretations, by downplaying the threat of nuclear power, except in extreme cases and in the hands of other nations.

The issues discussed on political shows were dramatized by entertainment shows like *The Twilight Zone*, *Space Patrol*, *Tales of Tomorrow*, *Watch Mr. Wizard*, and *Our Friend the Atom*, as entertainment shows could go further to visualize the implications of nuclear energy for their audiences. This made discussions of the atom more accessible to a wider audience and effectively increased the ability of television to play off real fears Americans had at the time by representing those fears on the screen. Each of these broadcasts contributed to the dualistic image of the atom by portraying scenarios in which America was on the receiving end of a nuclear strike, exploring the possible benefits and pitfalls of a world run on nuclear power, or explaining the potential that atomic energy had to advance not only fuel sources but medicine and agriculture as well. Entertainment television was having the same conversation with its audience as political broadcasts, it was simply doing so in a more user friendly and evocative manner.

The peaceful and destructive representations dominated discussions of the atom across a wide range of shows, most with large viewership. This essentially ensured that if the American public was watching something about the atom on television they were either being exposed to a narrative which reassured them of America's benevolent use of nuclear technology, or the frightening possibility of an outsider wielding the atom's destructive potential against the United States. This is significant because it helps us understand how television reflected and shaped cultural fears and hopes. Even though these representations were not so frequent that American audiences were faced with them every day, they appeared in such broad types of television broadcasting and even within schools that it was nearly guaranteed this message was seen by a vast majority of the American population.

While it is unlikely that these shows radically altered many people's opinions on how the atom should be used, they certainly perpetuated the idea that there were only two ways in which the atom could be discussed. During the Cold War, every day could have been the day the Soviets launched a nuclear strike on the U.S.; these fears would have been well-known and well-discussed among the American public who could cling to either the strength of deterrence or Eisenhower's hope for peace. Television came to reflect these fears and hopes in some of the most popular shows of the era. It is important for future research to consider the validity of television as a source for contextualizing the hopes, fears, and goals of the population it serves. By looking at television scholars can better understand what was important to its viewers, or what was important to producers for viewers to see.

Notes

1. Dwight Eisenhower, "Atoms for Peace: Address by Mr. Dwight D. Eisenhower, President of the United States of America, to the 470th Plenary Meeting of the United Nations General Assembly," (speech, New York, NY, December 8, 1953), International Atomic Energy Agency, https://www.iaea.org/newscenter/multimedia/videos/atoms-peace-speech, 00:18:23.
2. Chester E. Holifield, "The Birth of the Peaceful Atom," in *Atoms for Peace: An Analysis After Thirty Years,* ed. Joseph F. Pilat, Robert E. Pendley, and Charles K. Ebinger (Milton: Routledge, 2019), 53.
3. Michael Krepon, *Winning and Losing the Nuclear Peace: The Rise, Demise, and Revival of Arms Control,* (Redwood City: Stanford University Press, 2021), 41. https://doi-org.ezproxy.lib.vt.edu/10.1515/9781503629615
4. ---, "Soviet Tests," *American Experience,* PBS.org, , last accessed April 6, 2023. https://www.pbs.org/wgbh/americanexperience/features/bomb-soviet-tests/
5. Ernie Tretkoff "December 1938: Discovery of Nuclear Fission," *This Month in Physics History* 16, no. 11, Alan Chodos and Jennifer Ouellette, eds, (December, 2007). https://www.aps.org/publications/apsnews/200712/physicshistory.cfm#:~:text=December 1938: Discovery of Nuclear Fission&text=In December 1938, over Christmas,lead to the atomic bomb
6. Jacob Darwin Hamblin, *The Wretched Atom: America's Global Gamble with Peaceful Nuclear Technology* (New York: Oxford University Press, 2021). https://doi-org.ezproxy.lib.vt.edu/10.1093/oso/9780197526903.001.0001; Helen Anne Curry, *Evolution Made to Order: Plant Breeding and Technological Innovation in Twentieth-Century* (Chicago: University of Chicago Press, 2016). https://doi-org.ezproxy.lib.vt.edu/10.7208/chicago/9780226390116.001.0001; Angela N. H. Creager, *Life Atomic: A History of Radioisotopes in Science and Medicine,* (Chicago: University of Chicago Press, 2013). https://doi-org.ezproxy.lib.vt.edu/10.7208/chicago/9780226017945.001.0001.
7. Victoria E. Johnson, "The Classic Network Era in Television: 1950s-1970s," in *A Companion to the History of American Broadcasting,* ed. Aniko Bodroghkozy (New Jersey: John Wiley & Sons, Inc., 2018), 1.
8. Cynthia J Miller and A. Bowdoin Van Riper, *1950's "Rocketman" TV Series and Their Fans: Cadets, Rangers, and Junior Space Men,* (New York: Palgrave Macmillan, 2012), 165-176; Rodney Hill, "Mapping The Twilight Zone's Cultural and Mythological Terrain," in *The Essential Science Fiction Television Reader,* ed., J.P. Telotte (Kentucky: The University Press of Kentucky, 2008), 121-122.
9. Holifield, "The Birth of the Peaceful Atom," 56.
10. See, for example, "Duck and Cover," created by the Federal Civil Defense Administration, first aired in 1951. https://www.c-span.org/video/?524420-1/atomic-bomb-safety-students.
11. Eisenhower, "Atoms for Peace," 00:06:02.

12. Eisenhower, "Atoms for Peace," 00:06:55.
13. "Brief Chronology," Biographical Overview, National Library of Medicine, date accessed April 25, 2023, https://profiles.nlm.nih.gov/spotlight/mm/feature/biographical-overview.
14. Betty Duker, *Meet the Press,* aired May 11, 1958, NBC, https://video-alexanderstreet-com.ezproxy.lib.vt.edu/watch/meet-the-press-may-11-1958.
15. *Meet the Press,* May 11: 00:08:15.
16. *Meet the Press,* May 11: 00:08:30.
17. "Duck and Cover," 00:02:37.
18. ---, "Announcing the Bombing of Hiroshima," *American Experience,* PBS.org. accessed Apr. 6, 2023. https://www.pbs.org/wgbh/americanexperience/features/truman-hiroshima/.
19. "Atomic Energy Commission," Glossary, United States Nuclear Regulatory Commission, last modified March 09, 2021, https://www.nrc.gov/reading-rm/basic-ref/glossary/atomic-energy-commission.html.
20. Eisenhower, "Atoms for Peace," 00:21:19.
21. Eisenhower, "Atoms for Peace," 00:22:35.
22. *Chronoscope* was a fifteen-minute talk show that aired from 11:00 pm to 11:15 pm on CBS from 1951-1955. LeSueur was also known for his time as a CBS war correspondent during World War II.
23. *Chronoscope*, December 9, 1953, anonymous Columbia Broadcasting System, 1953, https://video.alexanderstreet.com/watch/rep-carl-hinshaw-r-ca: 3:19
24. *Ibid.*
25. *Chronoscope*, December 9: 4:00.
26. *Chronoscope*, December 9: 6:00; *Our Friend the Atom*, directed by Hamilton S. Luske, aired January 23, 1957, ABC, https://www.youtube.com/watch?v=pkwadgJORFM, 00:39:00.
27. *Chronoscope*, December 9: 6:00.
28. *Meet the Press,* May 11, 00:00:35; Robert Kleidman, *Organizing for Peace: Neutrality, the Test Ban, and the Freeze* (version 1st ed.), 1st ed, Syracuse Studies on Peace and Conflict Resolution (Syracuse, N.Y.: Syracuse University Press, 1993), 1, https://archive.org/details/organizingforpea0000klei/page/n1/mode/2up.
29. *Meet the Press,* May 11: 6:20.
30. *Meet the Press,* May 11: 9:35.
31. Creager, *Life Atomic,* 9.
32. Creager, *Life Atomic,* 14.
33. Creager, *Life Atomic,* 14; 260.
34. Curry, *Evolution Made to Order,* 180; John E. Gray, "The Ambitious Rush Toward Atomic Power," in *Atoms for Peace: An Analysis After Thirty Years,* ed.

Joseph F. Pilat, Robert E. Pendley, and Charles K. Ebinger (Milton: Routledge, 2019), 126-127; Scott Kaufman, *Project Plowshare: The Peaceful Use of Nuclear Explosives in Cold War America* (Cornell University Press, 2013), 13-15, http://www.jstor.org/stable/10.7591/j.cttn34r4.

35. Creager, *Life Atomic,* 11.
36. Gray, "The Ambitious Rush Toward Atomic Power," 126-127.
37. Miller and Bowdoin Van Riper, *1950's "Rocketman,"* 163.
38. *Ibid.*
39. Miller and Bowdoin Van Riper, *1950's "Rocketman,"*166-167.
40. Miller and Bowdoin Van Riper, *1950's "Rocketman,"* 165.
41. Ibid.
42. *Ibid.*
43. *Space Patrol,* season 4, episode 7, "The Deadly Radiation Chamber," directed by Dick Darley, aired February 13, 1954. https://tubitv.com/tv-shows/708677/s04-e07-the-deadly-radiation-chamber?start=true; *Space Patrol,* season 4, episode 8, "The Plot in the Atomic Plant," directed by Dick Darley, aired February 20, 1954. https://tubitv.com/tv-shows/708670/s04-e08-the-plot-in-the-atomic-plant?start=true.
44. "The Deadly Radiation Chamber," 00:06:00.
45. "The Deadly Radiation Chamber," 00:17:16.
46. "The Deadly Radiation Chamber," 00:19:00.
47. "The Deadly Radiation Chamber," 00:24:00.
48. *Tales of Tomorrow,* season 1, episode 7, "Sneak Attack," directed by Charles S. Dubin, aired December 7, 1951. https://www.amazon.com/Frankenstein/dp/B078YC3DWD/ref=sr_1_1?crid=2N2JJLWL9JE5K&keywords=tales of tomorrow&qid=1677469317&s=instant-video&sprefix=tales of tomorrow,instant-video,83&sr=1-1, 00:08:00.
49. "Sneak Attack," 00:12:26.
50. "Sneak Attack," 00:14:58.
51. "Sneak Attack," 00:15:50.
52. "Sneak Attack," 00:18:40.
53. "Sneak Attack," 0:22:40.
54. "Sneak Attack," 00:16:55.
55. "Sneak Attack," 00:17:55.
56. Hill, "Mapping *The Twilight Zone's* Cultural and Mythological Terrain," 121-122.
57. *The Twilight Zone*, season 1, episode 8, "Time Enough at Last," created by Rod Serling, aired November 20, 1959. https://pluto.tv/en/on-demand/series/5ef0d81d146bc5001ae9f7d0/season/1/episode/61c506c6f72e7b001a68d074, 00:30:00.

58. "Time Enough at Last," 00:09:57.

59. "Time Enough at Last," 00:10:24.

60. "Time Enough at Last," 00:10:34.

61. "Time Enough at Last," 00:11:33.

62. "Time Enough at Last," 00:23:00.

63. Ingrid Ockert, "Science Television in the Sputnik Age," *Physics Today* (September 22, 2017).

64. *Ibid.*

65. *Watch Mr. Wizard*, season 6, episode 38, "Atomic Energy," created by Don Herbert, aired 1953, https://www.youtube.com/watch?v=Ln7DcQhqBJY, 00:01:14.

66. *Watch Mr. Wizard,* "Atomic Energy," 00:09:40–00:10:00.

67. Ockert, "Science Television in the Sputnik Age."

68. *Ibid.*

69. *Our Friend the Atom,* 00:01:22–00:02:49.

70. *Our Friend the Atom,* 00:04:45.

71. *Our Friend the Atom*, 00:07:21.

72. *Our Friend the Atom,* 00:34:40.

73. *Our Friend the Atom,* 00:37:00.

74. *Our Friend the Atom,* 00:42:14–00:47:14.

75. *Our Friend the Atom*, 00:41:20.

Bombs Over the Chess Board

JASON TOY

In July 1972, the eccentric and unpredictable chess Grandmaster Bobby Fischer set off an international incident that sent the chess world into disarray. The United States Chess Champion backed out of his World Championship match with reigning world champion Boris Spassky of the Soviet Union, set to take place in the first week of July in Reykjavik, Iceland. The potential implications of declining a match with the Soviet champion in the context of the Cold War Détente led the United States national security advisor Henry Kissinger getting involved to rectify the situation. On July 3, Kissinger called Fischer and introduced himself as "the worst chess player in the world calling the best chess player in the world."[1] He stressed to Fischer that the country's prestige was at stake and convinced the chess champion to fly to Iceland the next day.

In the context of the Cold War, sports became another facet of international relations. Since 1948, the United States and the Soviet Union had been engaged in a political, economic, and military competition to demonstrate the superiority of their respective systems. Proxy conflicts, such as those sparked by the United States' containment policy during the early portion of the Cold War (in Greece and Italy after World War II, for example, or the Vietnam War—see Danielle Schuhmacher's contribution to this volume), aimed to limit the spread of communism worldwide. As the Cold War continued, proxy conflicts were centered in other types of competition as well. As early as the first postwar Olympics in 1948, international sports competitions became another field of battle: countries could prove their superiority over their rivals, and sports victories were often used to boost national pride and prestige.

The 1972 World Chess Championship match was particularly significant because it pitted two individuals against each other in a game of intellect and strategy, in which the players became representatives of their respective ideological systems. This chapter examines how the media, and television in particular, constructed the 1972 World Championship match as a proxy conflict between the two superpowers. Sports media seized upon chess as another front in the symbolic competition between the US and the USSR. But chess, long ignored by sports media, fit uncomfortably into established tropes of Cold War conflict. Media coverage portrayed Bobby Fischer, an unconventional and impulsive player and reluctant celebrity, as the embodiment of American individualism and the free market economy. On the other hand, international coverage defined Boris Spassky, a stoic and disciplined player, as a representative of the Soviet Union's collectivist and state-controlled system. The match, therefore, became a battle not just between two individuals but between two ideological systems.

My research explores how television used the 1972 World Chess Championship match to construct a proxy conflict between the United States and the Soviet Union. This chapter first establishes what has already been written about the match and then provides critical context pertaining to chess and television history. The majority of the chapter will pertain to the match itself, with particular focus on the narrative created by television coverage, and how the Cold War rivalry narrative affected the sport and the people involved.

HISTORIOGRAPHY

This chapter brings together research from sports history, the cultural history of the Cold War, and media studies to show how television fits chess into the larger geopolitical rivalry of the Cold War. There is ample scholarship from sports history addressing the individual games and the life of Bobby Fischer. The literature, including David Edmond's *Bobby Fischer Goes to War* and Brad Darrach's *Bobby Fischer vs. the Rest of the World*, conforms to the media narrative created around Bobby Fischer in the early 1970s. These accounts focus on how the proxy conflict affected both him and the world: the 1972 championship was a global phenomenon, and Fischer was a complex Cold War icon. Edmond's book covers the political, social, and cultural context surrounding the match, while Darrach's book gives an early insight into Bobby Fischer's life and provides a more in-depth analysis of the games. Frank Brady's *Endgame: Bobby Fischer's Remarkable Rise and Fall*, by contrast, is a psychological history of the man. Brady used journals, memoirs and personal accounts by Fischer, or his close family members and friends, medical professionals and psychologists to try and get an understanding of Bobby's mental state/condition during certain events in his life. For Brady, Fischer's brilliance as a chess player was inextricably linked to his personal demons, and his legacy is a complex and multifaceted one. George Chressanthis investigated the United States Chess Federation in terms of its popularity, demand, membership and prize earnings. Finally, Gary Alan Fine has helped me put the match into the historiographical context of chess history, comparing the match's significance to other chess events.[2]

This chapter also engages the cultural history of the Cold War and the role of symbolic competition, particularly the superpower rivalry in the realm of sports and athletics. It highlights how these contests projected national power and influence, as well as promoted the values and ideologies of the competing nations. The historiography of the Cold War is comprised no longer of a straightforward political history of events but has been expanded in explorations of the social, cultural, and technological changes that took place during the time of increased international tension between the United States and the Soviet Union. Scholars such as Thomas Hunt and Dennis Coates have shown that sports was central to Cold War competition. Hunt, for example, examined the significance of American sports policy under President Lyndon Johnson's administration. In the book *White King and Red Queen,* author Daniel Johnson tells of the deep symbolism and psychological warfare that characterized chess games between the U.S. and the Soviet Union during the Cold War.[3]

Finally, I draw upon a media studies scholarship and important primary sources to show how the medium of television reframed sports as another proxy conflict of the Cold War. Travis Vogan explored the ABC Worldwide Sports television broadcast to trace efforts to bring attention to obscure sports and forms of competition. A New York Times report on the passing of Shelby Lyman, one of the commentators of the match, gives a behind the scenes account of broadcasting decisions for the match in the United States. Finally, interviews with Bobby Fischer and Boris Spassky on The Dick Cavett Show, The Tonight Show Starring Johnny Carson, 60 Minutes, and AP News provide firsthand evidence of how western television packaged chess and portrayed it to western and American audiences.[4]

A QUIET GAME

Chess is an ancient game that has been played for millenia. It is a game of strategy and tactics with the goal of capturing the opponent's King, a position known as checkmate. Two players take turns moving their pieces; each piece moves in its own unique pattern across a sixty-four square board. Before the 1972 World Chess Championship, chess was already a well-established sport with a rich history and culture, but it primarily found its place in people's homes, community halls, or even outdoor parks. The general public primarily understood chess as an intellectual pursuit that appealed to enthusiasts.

Although millions of people already played chess, it had not attracted the attention of burgeoning sports media. Chess did not operate like more traditional "sports" such as baseball or football, and the professional scene was unattractive to sports producers of the new television medium. Professional classical games (the main form of competitive chess) were often over 180 minutes of two people silently moving their pieces on the board; marketing a quiet board game on television was quite difficult. Professional tournament matches were also long, comprising several games to determine the winner. Typically, professional games were played until a player earned four or six points, with draws giving both competitors half-a-point and wins awarding the winner a full point. The conditions of a long chess tournament (silence and an indeterminate timetable of play) paled in comparison to the exciting live atmospheres built up around traditional sporting events.

Professional chess also differed from more established sports in that its ranking and scoring system was complicated and inscrutable to the lay audience. The International Chess Federation, best known by their French acronym FIDE, regulated the ranking of players through the Elo rating system. The system calculated the relative strength of chess players based on their performance in rated games. Developed by the Hungarian-American physicist Arpad Elo, the system has since become the most widely used rating system in chess.[5] The scale is zero to infinity; players with an elo over 2500 achieve the highest title of "Grandmaster."

This unique way of determining the best has deterred general audiences from following the sport. First, the Elo system is based on the performance of individual players in rated games, whereas many traditional sports have team-based rating systems that take into account various factors such as team composition, strategy, and tactics. In chess, a player's rating is solely determined by their individual performance in games against other players. Second, the Elo system uses a mathematical formula to calculate ratings, which may involve calculations and adjustments that are not immediately transparent or intuitive to those who are not familiar with the underlying mathematical principles. Third, the Elo system relies heavily on historical data and the ratings of other players, which can sometimes result in unexpected changes in ratings based on relative performance. Finally, the Elo system is continuously updated after each rated game, which means that ratings can fluctuate frequently, making it challenging for casual observers to keep track of changes.

Among enthusiasts, chess had always been seen as a gentleman's game, a very respectful affair, which further undermined its appeal to mid-century media. Popular athletes of the time, such as the boxer Muhammed Ali, attracted attention to their sports because of their eccentric, larger-than-life personalities and fun sound-bites. Chess had few eccentric personalities that attracted large audiences and broadcasting companies to the games, which helps explain why chess was not widely covered in the mainstream sports media on television in the United States during the Cold War. Instead, chess was primarily covered on radio and in the print media: chess radio stations carried live game commentary, while the print media covered post-game analysis. Only in the Soviet Union were professional games and tournaments regularly broadcasted on television.

Finally, the stakes for chess seemed low. Chess tournaments were often held in small venues with limited audiences, and the prize money was relatively paltry compared to other professional sports. Chess tournaments were usually funded by private investors, national chess federations, or corporate sponsorships. Private investment was new, emerging after the 1972 World Champion as the game had only just reached the interest of eccentric millionaires. The sole national chess federation that committed large prize funds was that of the Soviet Union; there the organization was fully backed by the national government. Sponsors were usually made up of local businesses or companies where the tournament was located, and the sport did not attract large brands for funding. Prize money for professional tournaments accumulated in the low thousands. Boris Spassky, for example, won just $1,400 for winning the 1969 World Chess Championship, the most prestigious and well-funded tournament in the sport.[6]

The match between Fischer and Spassky changed this perception of chess. The intense media coverage of the match brought the game to the attention of millions of people around the world, and it introduced a new level of drama and excitement to the sport. The media thrust Fischer, in particular, into the spotlight, and he captured the imagination of the public with his unpredictable behavior and an eccentric personality that he had developed throughout his chess career. Moreover, observers anticipating the match framed it as another battle between the two superpowers, and once the match was scheduled, television media covered the rivalry extensively and constructed the match as a proxy conflict of the Cold War.

RIVALRY

The Soviet Union was a dominant force in chess leading up to the 1972 World Chess Championship, thanks to government support, strong coaching programs, and a culture that valued intellectual pursuits. After World War II, the Soviet government had heavily invested in chess programs, funding coaching and training, and supporting national and international tournaments. The Soviet chess federation had developed a comprehensive coaching system that produced many world-class players. The culture of the Soviet Union valued chess as a noble pursuit and celebrated successful players as heroes.[7] The Soviets valued intellectual competition and traditional sports the same. This resulted in the dominance of Soviet players in international chess, with numerous world championships won and major tournaments hosted in the country.

The United States lagged behind the Soviet Union in the world of chess prior to the 1972 World Chess Championship, mainly due to a lack of centralized support and coaching programs. Unlike the Soviet Union, which heavily invested in chess development, the U.S. government did not provide significant funding or support for chess. Before 1972, the U.S. government did not believe that intellectual competitions such as chess were viable or important arenas of the Cold War that were worthy of investment. Additionally, the smaller pool of chess players in the U.S. limited opportunities for American players to compete against top-level competition and improve their skills. These factors posed challenges for American chess players in competing at the highest international levels.[8]

Bobby Fischer and Boris Spassky were the two best chess players in the lead up to the 1972 World Championship. Fischer, representing the United States, and Spassky, representing the Soviet Union, were both renowned players who captivated the chess world with their unique playing styles, contrasting backgrounds, and intense competition on and off the chessboard. Fischer, born in Chicago in 1943, was a child prodigy who showed exceptional talent and dedication to chess from a young age. Bobby had learned the game from his sister, Joan, and obsessively played with the set by himself. Worried about her son's reclusive behavior, Bobby's mother Regina tried to post an ad in the Brooklyn Eagle newspaper to find playmates for her son. The paper rejected her ad, but Brooklyn Eagle journalist Hermann Helms wrote back to Regina recommending that Bobby play at the Brooklyn Chess Club under Carmine Nigro.[9] Bobby was so obsessed with the game that he pestered his mother every day to play at the chess club or at Manhattan's Washington Square Park. Regina became so concerned that chess was ruining Bobby that she took him to see a psychiatrist about his obsession. The doctor advised Regina that "there are worse things to be obsessed over than chess."[10]

Fischer was known for his meticulous preparation, innovative ideas, and aggressive playing style. Fischer was also known for his uncompromising nature and fierce determination to win, which earned him a reputation as a formidable opponent on the chessboard. At the age of fifteen Fischer became the youngest person to earn the title of Grandmaster.[11] Despite his outstanding play on the board, Fischer's reclusiveness, irrational behavior, and erratic demands for money and control over playing conditions had garnered him a bad reputation within the chess community.

Boris Spassky, on the other hand, was another child prodigy who was known for his calm demeanor during games. Born in Leningrad (now St. Petersburg) in 1937, Spassky first learned chess at the age of five from his parents while fleeing the Siege of Leningrad during World War II. In 1947, at the age of ten, Spassky defeated the reigning Soviet champion Mikhail Botvinnik in an exhibition game that drew the attention of the Soviet Chess federation. In 1956, Spassky became the youngest player to achieve the "Grandmaster" title in chess, a record that Bobby Fischer broke just two years later.[12] In 1969, Spassky challenged reigning Soviet-Armeninan world champion Tigron Petrosian for the World Championship in Moscow. Spassky won by two points, reaching new heights in the world of chess, while also continuing the winning streak of Soviet chess champions.[13]

Spassky was a highly versatile player, capable of adapting to different opponents and situations. He was a product of the Soviet chess system, known for its emphasis on coaching, training, and state support for chess players. Spassky had all the Soviet Grandmasters to practice against daily, access to a coaching staff that had constructed a strict training regiment for him, and the financial support of the Soviet government. His success on the chessboard can be (and in the press it was) attributed not only to his natural talent but also to the rigorous training and support he received from the Soviet chess system.

In the early 1960s, as both players ascended to the top ranks of the chess world, they increasingly met at the chessboard and a rivalry emerged. In 1960 they had their first match at the Mar del Plata Tournament in Argentina. Spassky defeated Fischer during the round-robin portion of the competition, which meant that Fischer had only faced Spassky once when the two players initially tied in the event.[14] Fischer's attitudes regarding Soviet chess players had shifted by 1962, which led to a rivalry between him and the Soviet Union. The Stockholm Interzonal Chess Tournament determined the qualifiers for the Candidates Tournament, the next step towards challenging for the World Chess Championship. However, the tournament was marred by controversy. In a *Sports Illustrated* article reporting on the tournament, Fischer made allegations of collusion and cheating against Soviet players, particularly Paul Keres, Efim Geller and Mark Taimanov. Fischer accused them of pre-arranging draws to ensure their qualification at the expense of other players, including himself. FIDE investigated the matter and did not find enough evidence to punish the Soviet players.[15] Fisher's accusations compromised his relationships with Soviet chess players, even with players who did not compete in the tournament, such as Spassky. Fischer's lifelong enmity with the Soviet chess players began as a result of the controversy.

Four years later, Fischer and Spassky met for a second time at the 1966 Piatigorsky Cup held in Santa Monica, California. The tournament attracted the top players in the world, most of them Soviet. Spassky and Fischer faced each other twice in this tournament, and both games ended in draws. Spassky edged out Fischer by a half point, marking Fischer's first tournament loss to Spassky.[16] That same year they met at the Chess Olympiad in Siegen, Germany; this was their last meeting before the World Championship game in 1972. Placing second in the tournament, Fischer beat all the other Soviet Chess Grandmasters at the tournament, but he lost to Spassky again.[17] Fischer's past interactions with the Soviet chess team, as well as his rivalry with (and 0-3 record against) Spassky, gave the media enough material to build interest in their next meeting at the World Championship, the match that became another Cold War proxy conflict.

ROAD TO REYKJAVIK

Bobby Fischer took an eighteen month break from chess competitions from 1968-1969. But in order to qualify to challenge for the World Championship title, one must compete in a series of qualifying tournaments: the zonal, interzonal, and the Candidates tournament determine the best player to challenge the reigning champion for the World Championship. Despite possessing the highest chess rating, Fischer had refused to compete in the zonal tournament during his break in 1969, which would normally disqualify him from competing in the next two tournaments. Confronted with Fischer's missing zonal tournament win, the United States Chess Federation convinced two grandmasters, Hungarian Paul Benko and American William Lombardy, to let Bobby compete in the interzonal. They argued that that Fischer was the only person who had a chance of beating Spassky, and even paid them off to left him compete.[18]

In November and December 1970, the Interzonal competition occurred in Palma de Mallorca, Spain, with the top six finishers moving on to the Candidates matches. The Candidates competitions took place in 1971, with the winner earning the chance to take on Boris Spassky, the current world champion. The matches were held in various locations around the world, including Vancouver, Canada, and Belgrade, Yugoslavia. The broadcasting of these tournaments continued to be quite limited compared to the 1972 championship match, but it still played a crucial role in expanding the reach of the game. In the United States, the matches were still covered by radio, newspapers and magazines. In other parts of the world, television played a more significant role in broadcasting the matches. The Soviet Union continued to broadcast the games on national television. Alongside television broadcasts, the Soviets had dedicated their radio broadcast during the tournament to track Bobby Fischer's progress. The Soviets even reported on his daily activities, such as when he left his hotel and who accompanied him, daily on the radio.[19]

The tournament featured many of the top players in the world at the time, including Soviet grandmasters Efim Geller, Mark Taimanov, and Vasily Smyslov, as well as American grandmaster Samuel Reshevsky. These players were all vying for a chance to move on to the next stage of the World Chess Championship cycle, and the competition was fierce. The matches were played over a two-month period, with each player competing against several others in a round-robin format. Despite drawing and losing his first few games, Bobby Fischer won the tournament and went on to play in the Candidates tournament the following year along with the five players who placed below him in the tournament. They were joined by Soviet Grandmasters Tigran Petrosian, a former world champion, and Viktor Korchnoi, who had advanced as the loser of the 1968 Candidates match against Spassky.[20]

As Fischer's success gained greater attention, he found new audiences on American television. In 1971 Dick Cavett hosted him on the ABC talk show *The Dick Cavett Show*.[21] At the time of the interview, Bobby Fischer was preparing for the Candidates matches that would determine who would face Boris Spassky for the World Chess Championship. He was an introvert, notoriously private, and he avoided the press, making his appearance on the show a major event. Fischer was already known for his eccentricities and reclusiveness in the chess community, but the interview provided a rare opportunity for the public to see Fischer speak candidly and express his views on a range of topics. During the interview, Fischer displayed his trademark intensity and determination, discussing his rigorous training regimen and unwavering focus on the game of chess. The interview drew a larger television audience than Fischer had ever been exposed to and increased his popularity, helping to build hype around his possible upcoming World Championship match.

The majority of Cavett's questions were lighthearted, designed to introduce Fischer and competitive chess to the average American. Cavett asked Fischer where he came from, about his lifestyle, and how the world of professional chess operated. In the latter part of the interview, Cavett was the first person in television to present Fischer's rivalry with the Soviets in terms of the larger geopolitical conflict. Fischer's responses emphasized his patriotism, which was new even to those who knew him from the international chess community. He expressed his disdain from the Soviet chess establishment and iterated his belief that they were colluding against him.[22] The way Cavett structured the interview steered the narrative surrounding Bobby Fischer from a reclusive selfish person to determined patriot preparing to represent his country.

The Candidates tournament of 1971, held in various locations around the world, including Vancouver, Palma de Mallorca, and Buenos Aires, was the last hurdle to the world championship. The top six players from the interzonal tournament, the former World Champion, and the 1968 Candidates finalist competed to advance to the World Championship. The best of ten games format was used for the quarterfinal and semifinal matches and the final was a best of twelve. The match followed standard FIDE scoring, half a point to both players for ties, one full point to the player that wins.

Bobby drew Soviet Mark Taimonov as his quarterfinal match. After placing first in the interzonal, Bobby's confidence was at an all-time high, and he assured the media that he would win this match as well. The Soviet Union's Communist party's daily newspaper, *Pravda*, was pessimistic of Taimanov's chances as he was, by all measures, the weakest Soviet Grandmaster participating in the tournament.[23] Indeed, Fischer beat Mark Taimonov and Danish Grandmaster Bent Larsen in the quarter- and semifinals in both clean six-zero sweeps.

This tournament generated little public interest in the United States, but the President was watching. President Richard Nixon sent Bobby Fischer a letter before his Candidates Final match against Soviet Grandmaster and former world Champion Tigran Petrosian to express his admiration and offer the well wishes of the nation. In the letter, Nixon wrote:

> I wanted to add my personal congratulations to the many you have already received. Your string of nineteen consecutive victories in world-class competition is unprecedented, and you have every reason to take great satisfaction in your superb achievement. As you prepare to meet the winner of the Petrosian-Korchnoi matches, you may be certain that your fellow citizens will be cheering you on. Good luck![24]

The Nixon letter perhaps jinxed Fischer. After drawing the first game against Petrosian, Fischer lost game two, ending a 20-game winning streak. Yet Fischer still beat Petrosian with a final score of 6.5 to 2.5, becoming the challenger for the 1972 Chess World Championship.

On April 9, 1972, just a few months before the now highly anticipated 1972 World Chess Championship match between Bobby Fischer and Boris Spassky, Fischer appeared on the American television news program *60 Minutes*. The interview, with lead reporter Mike Wallace, provided a rare glimpse into Fischer's personality and his thoughts on the upcoming match. At the start of the interview, the show ran clips of Fischer's physical and mental training regimen, showing Fischer doing situps and practicing chess while reading a book. These clips presented professional chess as a sport that demanded both physical endurance and mental fortitude, a counter-narrative for the prime time American audience, which generally believed that chess was not a sport.

In the face-to-face interview with Fischer, Wallace wasted no time in asking Fischer about his rivalry with the Soviets. Fischer took the chance to explain his history with the Soviet Chess Federation. In particular, he defined the upcoming game not as a grudge match between individuals (him and Spassky,) but rather framed it as a grudge match between him and the entire Soviet Chess empire.[25] The interview was a significant moment in the lead-up to the 1972 World Chess Championship, as it provided a rare opportunity for the American public to hear from Fischer, who characterized his challenge as one of fighting the Soviets. It also helped to unravel Fischer's reputation as a complex and enigmatic figure.

WHERE'S BOBBY?

The match was set to take place in July, but as negotiations developed in the months prior to the match, Fischer had not yet signed any documents guaranteeing that he would play in the world championship. His main issue was money. Fischer believed that the biggest chess tournament in the world should have a larger prize pool. Moreover, due to the shift towards broadcasting chess events, he argued that both he and Spassky should get more money from the media rights to the games. Just days before the match it appeared that the financial arrangements had been resolved. Both participants would earn thirty percent of the television and film rights; in addition, the winner would receive $78,125 with the loser receiving $46,875. However, Fischer always followed the policy of neither signing, confirming, nor agreeing to anything. He now contended that the pot also should include thirty percent of the gate earnings, which were expected to total $250,000. The Icelanders hosting the event objected because they needed this cash to fund their expenses; the exhibition hall's capacity was approximately 2,500 with no room for expansion.[26]

Although not all of Fischer's financial demands were met, he scheduled a flight to Iceland on 28 June. When he arrived at John F. Kennedy airport in New York, he was swarmed by cameramen trying to capture his historic departure. Anxious, Fischer bolted away from the media, jumped into the first cab available, and fled the airport.[27] Fischer's disappearance was widely covered in the media. The major American news networks ABC, NBC, and CBS, followed the story in their nightly news programs. Fischer hunkered down in the home of former US Chess Champion Anthony Saidy in Queens, New York City and released a new set of demands for him to play in the World Championship.[28] It included better conditions pertaining to money, living conditions, and media rights. The press soon discovered where Fischer was hiding, and they descended upon the residence. Outside of Saidy's house, a media encampment with cameras and trucks was set up in an effort to photograph Fischer or conduct an interview. In the meantime, the first match of the championship took place on 1 July. At the opening ceremony, the president of the Icelandic Chess Federation, Gudmundur Thórarinsson, refused to make a speech as he was embarrassed that the match had most likely collapsed. The minister of Icelandic culture Torfi Olafsson and Mayor of Reykjavik Geir Hallgrimsson made vague statements about the match. None of them noted Fischer's absence. Only Max Euwe, president of FIDE, addressed the Fischer situation saying "Mr. Fischer is not an easy man. But we should remember that he has lifted the level of world chess for all players."[29]

Behind the scenes, Icelandic officials scrambled to get Fischer to play. Prime minister of Iceland, Ólafur Jóhannesson, contacted the U.S. charge d'affaires Theodore Tremblay to see if the US government could help the situation. Tremblay followed up with a telegram to the Secretary of State William Rogers. In turn, the situation was presented to Henry Kissinger, then the United States Secretary of State. According to Kissinger's memoirs, he called Fischer on July 3, when Fischer expressed his concern about the prize fund and threatened not to play. Kissinger reportedly told Fischer that his absence would be a propaganda victory for the Soviet Union and urged him to reconsider.[30] Kissinger framed the significance of the match in terms of the Cold War rivalry between the United States and the Soviet Union. The 1972 match was the first time an American had competed in a World Chess Championship match since 1921, and Kissinger believed that Fischer might be the only chance the United States could beat the Soviets in a game that Soviet players had dominated since 1948.

In the end, a private citizen saved the day. James Slater, a wealthy British businessman and chess enthusiast, offered to double the prize fund to $250,000. Slater was introduced to Fischer by intermediaries: journalist Leonard Barden and Fischer attorney Paul Marshall.[31] With pressure from the press, Kissinger, and Slater, Fischer ultimately agreed to play. On the evening of July third, Fischer accompanied by Marshall, boarded a plane for Reykjavik.

THE MATCH OF THE CENTURY

Behind the scenes, preparations were underway for what promised to be a groundbreaking broadcast of the World Championship match. The match, held in Reykjavik, Iceland, was watched by millions of people around the world, and television played a crucial role in constructing a proxy conflict between the two superpowers. Preparing to broadcast the 1972 chess world championship was a complex operation. The actual filming and photography rights to the games were bought by a man named Chester Fox, a young up-and-coming filmmaker recommended to the Icelandic Chess federation (ICF) by Fischer's attorney Paul Marshall. Fox and the ICF stuck a deal with the intention of selling the film and still images to major media companies.[32]

Coverage of the 1972 chess world championship had a significant impact on the popularity of chess around the world. In the United States, all three major news networks—NBC, ABC, and CBS—sent news crews to Iceland.[33] To allow the moves from the game to be relayed back to analysts at home would could provide commentary in real time, their crews set up special telegraph connections to the United States. American Grandmaster Larry Evans hosted the program on ABC while Shelby Lyman, a Harvard dropout and former Sociology professor, did play-by-play analysis for PBS. The high-quality analysis and commentary helped to make the game more accessible to a wider audience. Lyman's broadcast attracted millions of American viewers and achieved the highest rating ever in public television.[34] Additionally, the match's coverage highlighted the intellectual and strategic nature of the game, which appealed to many people who had previously not been interested in chess.

The championship match followed a similar rule set as the Candidates matches. The defending champion needed twelve points to win the title; the challenger needed 12.5 to defeat him. Each player received half a point for ties and a complete point for victories. The first game of the match, played on July 11, 1972, was a dramatic and tense affair, with both players making bold moves in an effort to gain an early advantage. During the game, Fischer made a crucial error by moving his bishop to a vulnerable position where it might be quickly trapped, allowing Spassky to gain a significant advantage. The mistake was immediately noted by commentators, and the film captured the reactions of the players and the audience in the auditorium.

Fischer's mistake sent a chill through the audience and stunned the commentators. Spassky calmly took advantage of Fischer's error and pressed his advantage. Despite Fischer's mistake, the game continued for a couple more hours, with both players demonstrating impressive skill and resilience. In the end, Spassky emerged victorious, winning the game after fifty-six moves. The mistake was also widely discussed in the media, with newspapers and magazines around the world reporting on the game and analyzing Fischer's mistake. On the ABC broadcast, Grandmaster Larry Evans commented on the move saying: "Bobby saw six moves ahead here, when he made the move. He just didn't see seven moves ahead."[35]

The second game was scheduled for the next day but Fischer failed to appear at the board, resulting in a forfeit. This caused an uproar in the chess community and drew intense media attention. The reaction to Fischer's forfeit was swift and widespread. Many chess experts and commentators were stunned by Fischer's decision to forfeit the game. Frank Brady, founding editor of *Chess Life* magazine, relayed the situation live on Larry Evans' ABC broadcast. Brady told Evans that Fischer was particularly concerned about the cameras in the playing hall, which he felt were too distracting. According to Brady, Fischer believed that the cameras were making too much noise and that the sound was affecting his concentration.[36] Fischer's behavior, though shocking to contemporary viewers, was not entirely unexpected. He had a long history of demanding changes to the playing conditions and had withdrawn from tournaments and matches in the past if he felt that the conditions were not satisfactory. However, his decision to forfeit the second game of the world championship was a shock to many who expected greater commitment and sportsmanship in the context of this competition with the Soviet Union. Moreover, the television cameras were an integral part of filming the match, which in turn was central to emerging sports reporting and attempts to set this match in the wider context of Cold war competition. They had been installed in the playing hall in contractual agreement with Chester Fox, who had been Bobby Fischer's pick.

Fischer demanded that for the championship to continue, subsequent games must be played away from the cameras. The meeting between FIDE, Chester Fox, and the representation of the two players, was tense. Fox had paid an unknown amount of money for the rights and was extremely upset, mainly with Fischer, who directly jeopardized Fox's deal. All parties came to an agreement that Game Three would be played in the "ping pong," a back room behind the auditorium, with no audience or cameras.[37] The move to the ping pong room caused a stir among the audience and the media, who were not accustomed to being excluded from such a high-profile event. More broadly, Fischer's move to the ping pong room for Game Three was seen by many as a reflection of the tense political climate of the Cold War, an example of the two superpowers jockeying for position and using every opportunity to score points against each other. The fact that a chess match could become the focus of such intense media coverage and political intrigue is a testament to the power of television in shaping public perceptions of world events.

Played on July 16, Game Three was the turning point of the championship match. When Fischer initially walked into the room, he noticed that a closed-circuit television camera had been installed without his knowledge. Enraged, Fischer demanded that the arbiter (an official who oversees games and makes sure the chess rules are obeyed) Lothar Schmid remove the camera from the room immediately. Spassky threatened to force the game back into the auditorium if Fischer continued to act so rashly. Schmid refused to cave in to either demand, grabbing both Fischer and Spassky and pushing them into their chairs to begin the game.[38] Fischer opened the game with the Benoni defense opening, a move he had never played before, which stunned Spassky. The opening is considered one of the most rare and aggressive in chess, almost guaranteeing that the game would not end in a draw.[39] The most important area of the board, the center, was quickly taken over by Fischer. Spassky was outplayed throughout the entire game, giving Fischer his first ever victory over the Soviet. The win gave Fischer a much-needed boost of confidence after his rocky start to the match.

After Game Three, it was Spassky's turn to complain about playing conditions. He complained to organizers that the lighting in the ping pong room was destabilizing his concentration. He also complained about the possibility of a bugged chair. Spassky believed that Fischer's team was listening in on his conversations with his team, and that they had tampered with the lights. The chair and lighting fixtures were checked by both FIDE officials and Spassky's team, each of which found nothing. The reaction of the television media to Spassky's complaints was mixed. Some commentators dismissed his complaints as excuses for his poor performance, while others took them seriously.[40] In response, Fischer agreed to resume play back in the main auditorium, but this time with an audience and no cameras. Fischer's win in Game Three gave him confidence during the next two games, and he was able to draw Game Four and win Game Five. The latter game was played in front of the crowd for the first time in the match. After his win in Game Five, the Icelandic crowd erupted, chatting "Bobby, Bobby" as he left the auditorium.[41]

Negotiations over film and photography continued behind the scenes while the games went on. Prior to Game Six, which was to be played on 23 July, Icelandic officials appealed to American network broadcaster ABC to broker a deal that would allow filming and photography that met Fischer's noise requirements. Multiple ABC executives, including Chet Forte and Lome Hassan, flew to Reykjavik to resolve the issues and promised Fischer that their cameras would film without any noise.[42] Game Six quickly entered into a complex tactical struggle, with both players exchanging blows in the center of the board. However, Fischer seized the initiative and, by move twenty-three, he had established a dominant position. Despite Spassky's best defense, Fischer continued to press forward, and on move forty-one, Spassky resigned the game. Larry Evans had few words to describe Game Six, commenting "it's just a beautiful game, I don't know what more can be said."[43] Moreover, television commentators including Shelby Lyman and Evans were clearly stunned when, as the crowd began to applaud, Spassky also rose from his seat and joined in the acclamation of Fischer's play. Spassky's gesture was a gentlemanly move that disrupted the narrative of individual rivalry—and presumed Soviet tactlessness—as a microcosm of US-Soviet competition.

Fischer took the lead at 3.5 to 2.5, and Spassky would never regain it. In Game Seven, Spassky fell behind early, falling in a piece deficit, but recovered to force a draw. ABC captured the game on film from a fixed point in the far back of the auditorium's balcony. ABC executive Some Hassan had made that call, believing that they had reached an agreement for the filming and photographing rights. But after the game Fischer learned from a radio news report that ABC had filmed the game without his knowledge and again became enraged. He demanded apologies from the

ABC executives and veto power over subsequent filming of the match. This pushed ABC over the edge: frustrated with Fischer, ABC President Roone Arledge announced the company's exit by telegram.[44] For his part, Chester Fox agreed to a settlement with the ICF before the last game on August 27. It covered losses he incurred as a result of Fischer's obstruction of the filming of the matches and included a provision that Fox would receive a portion of the earnings from the match.[45]

Spassky won just one of the remaining fourteen games, which included ten ties. Fischer won four of the remaining matches and the World Championship title with a score of 12.5-8.5 in the decisive twenty-first game on August 31. Despite his victory, Fischer's volatile behavior continued. After the game, Fischer stormed out of the auditorium and fled into the hills outside Rekjavik. He spent the next day in seclusion amongst sheep reading newspaper headlines reporting his victory. The television media's reaction to Fischer's win was overwhelmingly positive, with many commentators praising his brilliance on the board and his ability to outmaneuver his Soviet opponent. Moreover, they framed his victory as a triumph for American individualism and a testament to the superiority of the free market system over the Soviet Union's collectivist approach.

Boris Spassky received a cold welcome back in the Soviet Union, which *The Associated Press* referred to as the "anti-VIP" treatment. Uncharacteristically, he had to wait in line for his bags, complete documents for customs, then stand in line for passport control with regular members of the public. Rather than shuttling home in the official Chaika limousine, a shabby gray and blue bus was waiting for him. His bus stopped at every traffic light, a clear insult, when in his triumph he had traveled through Moscow as if he were Brezhnev. Moreover, Spassky was persecuted by his Soviet chess peers. Former World Champion Mikhail Botvinnik remarked in the Soviet press that Spassky's defeat was due to his inflated confidence. Vasili Smyslov, another former world champion, also scolded Spassky in the press.[46]

When Bobby Fischer returned to the United States, by contrast, he was greeted as an American hero. Television media ignited a fiery interest in Fischer and the rivalry between the United States and the Soviet Union, so this World Championship received attention unlike any other from the American market. Fischer's fame was at an all time high, and he made an appearance on the *Tonight Show with Johnny Carson.* The host, Johnny Carson, treated him in a manner that was very similar to what Dick Cavett had done before, posing lighthearted questions about Fischer's lifestyle and how professional chess operated. Carson cracked several jokes with Fischer, getting him to laugh several times, which made him appear more relatable and helped to dispel the arrogant persona that he had developed throughout his career.

Following up on the softball questions, Carson asked Fischer about the broadcasting and playing condition scandals that occurred during the match along with his relationship with Spassky. Fischer responded saying "I didn't try to psych Spassky . . . . he appreciates me because I get the best conditions, and because I fight for all the things, he just has to sit back and, uh, he knows that the best lighting and the best chess set and the best everything."[47] Similar to Fischer's other appearances on late-night television, Carson seized the opportunity to stoke the Cold War rivalry by asking Fischer for his thoughts on the state of Soviet chess after his triumph. In a shocking response, Fischer defended his biggest rival, saying that Spassky was being used as a scapegoat. According to Fischer, the Soviets claimed that Spassy lost, not because of the communist system, but rather because he did not prepare properly.[48] Thus, Fischer, who previously had characterized the Soviet grandmasters as cogs in a well-developed system of training and patronage, now attacked the Soviet chess system while simultaneously defending the skills and motives of his biggest rival. In his final late-night television appearance, Fischer expressed respect for Spassky, revising his part in the narrative of Cold War rivalry that had been in development since 1962, while the larger story of US-Soviet conflict remained intact.

CONCLUSION

The 1972 World Chess Championship match between Bobby Fischer and Boris Spassky was more than just a battle of intellect and strategy; it was a symbolic representation of the Cold War competition between the United States and the Soviet Union. The media, and television in particular, played a crucial role in constructing the match as a proxy conflict between the two superpowers. The media's portrayal of Fischer as an unconventional and impulsive player, and Spassky as a stoic and disciplined representative of the Soviet Union's collectivist and state-controlled system, played upon and reinforced established Cold War tropes. The coverage of the match became a powerful tool for nationalistic propaganda, used to boost national pride and prestige.

Moreover, the 1972 World Chess Championship match signaled a shift in the perception of chess as a sport. Chess had long been regarded as a niche sport, overshadowed by more popular and mainstream sports such as football, basketball, and baseball. However, the match brought chess to the forefront of global media attention, transforming it into a vehicle for political and cultural symbolism. The impact of the match on the sport of chess was significant, inspiring a new generation of chess players and fans. In the years following the match, chess became more accessible and mainstream in the United States and abroad, with the establishment of professional leagues, the creation of international tournaments, and even the development of computer chess programs. The 1972 World Chess Championship match was a defining moment in the history of chess, and its legacy continues to shape the sport to this day.

Notes

1. David Edmonds and John Eidinow, *Bobby Fischer Goes to War : How the Soviets Lost the Most Extraordinary Chess Match of All Time* (New York: HarperCollins, 2005), 98.
2. Edmonds and Eidinow, *Bobby Fischer Goes to War*; Brad Darrach, Bobby Fischer vs. the Rest of the World (New York: Stein and Day, 1974); Frank Brady, *Endgame : Bobby Fischer's Remarkable Rise and Fall—from America's Brightest Prodigy to the Edge of Madness* (New York: Crown Publishers, 2011); George A. Chressanthis, "The Demand for Chess in the United States, 1946-1990," *The American Economist* 38, no. 1 (1994): 17-26; and Gary Alan Fine, *Bobby Fischer's conquest of the world's championship: The psychology and tactics of the title match* (New York, NY: Ishi Press, 1973).
3. See, for example,Thomas M. Hunt and John Hoberman, *Drug Games the International Olympic Committee and the Politics of Doping, 1960-2008,* 1st ed, (Austin: University of Texas Press, 2011; Dennis C. Coates, "Weaponization of Sports: The Battle for World Influence through Sporting Success." *The Independent Review* 22, no. 2 (2017): 215–21. http://www.jstor.org/stable/26314817; and Daniel Johnson, *White King and Red Queen* (Boston: Houghton Mifflin, 2008).
4. Travis Vogan, *ABC Sports : The Rise and Fall of Network Sports Television* (Berkeley, CA: University of California Press, 2018); Dylan Loeb McLain, "Shelby Lyman, 82, Dies; Unlikely Star of a Fischer-Spassky Broadcast," *New York Times*, August 20, 2019; and Bobby Fischer, interview by Dick Cavett, *The Dick Cavett Show*, NBC, August 5, 1971; Bobby Fischer, interview by Mike Wallace, *60 Minutes*, CBS, April 9, 1972; and Bobby Fischer, interview by Johnny Carson, *The Tonight Show With Johnny Carson,* NBC, November 8, 1972.
5. Edmonds and Eidinow, *Bobby Fischer Goes to War*, 211.
6. Harold C. Schonberg, "Waiting for Bobby Fischer," *New York Times*, September 1973, 283.
7. Geir Kjetsaa, "Review of *Soviet Chess: Chess and Communism in the USSR*," *The Slavonic and East European Review* 45, no. 105 (1967): 575.
8. George A. Chressanthis, "The Demand for Chess in the United States, 1946-1990," *The American Economist* 38, no. 1 (1994): 17.
9. Edmonds and Eidinow, *Bobby Fischer Goes to War*, 5-6.
10. *Bobby Fischer Against The World*, directed by Liz Garbus (2011, United States: HBO Documentary Films), 12:44 to 12:55. https://www.youtube.com/watch?v=pavYCzvg2fM&list=PL6GShP_Z_nzgexzMiHla4p71ktpuvO4wX&index=8.
11. Edmonds and Eidinow, *Bobby Fischer Goes to War*, 10.
12. "Boris Spassky: The Life and Legacy of a Chess Grandmaster," *The Chess*

Journal online, Last Updated: April 26, 2023, https://www.chessjournal.com/boris-spassky/.

13. Spassky won the game 12.5 to 10.5 over Petrosian.
14. Edmonds and Eidinow, *Bobby Fischer Goes to War*, 12.
15. Edmonds and Eidinow, *Bobby Fischer Goes to War*, 13.
16. Frank Brady, *Endgame : Bobby Fischer's Remarkable Rise and Fall -- from America's Brightest Prodigy to the Edge of Madness.* (New York: Crown, 2011), 107.
17. Edmonds and Eidinow, Bobby Fischer Goes to War, 83-84.
18. Edmonds and Eidinow, *Bobby Fischer Goes to War*, 84-85.
19. Edmonds and Eidinow, *Bobby Fischer Goes to War*, 83.
20. Edmonds and Eidinow, *Bobby Fischer Goes to War*, 86.
21. Bobby Fischer, interview by Dick Cavett, *The Dick Cavett Show*, NBC, August 5, 1971.
22. Cavett, "Interview with Bobby Fischer," 1971.
23. Edmonds and Eidinow, *Bobby Fischer Goes to War*, 87-88.
24. Edmonds and Eidinow, *Bobby Fischer Goes to War*, 93.
25. Bobby Fischer, interview by Mike Wallace, *60 Minutes*, CBS, April 9, 1972.
26. Edmonds and Eidinow, *Bobby Fischer Goes to War*, 134.
27. Edmonds and Eidinow, *Bobby Fischer Goes to War*, 135.
28. Edmonds and Eidinow, *Bobby Fischer Goes to War*, 135.
29. Edmonds and Eidinow, *Bobby Fischer Goes to War*, 139.
30. *Bobby Fischer Against The World*, 27:18 to 27:55.
31. Edmonds and Eidinow, *Bobby Fischer Goes to War*, 148.
32. Edmonds and Eidinow, *Bobby Fischer Goes to War*, 162-163.
33. Edmonds and Eidinow, *Bobby Fischer Goes to War*, 226.
34. Edmonds and Eidinow, *Bobby Fischer Goes to War*, 226
35. Bobby Fischer Against The World, 38:55 to 39:00.
36. *Bobby Fischer Against The World*, 46:03 to 46:20.
37. Edmonds and Eidinow, *Bobby Fischer Goes to War*, 179.
38. Edmonds and Eidinow, *Bobby Fischer Goes to War*, 180-181.
39. *Bobby Fischer Against The World*, 48:03 to 48:22.
40. *Bobby Fischer Against The World*, 52:07 to 54:07.
41. Edmonds and Eidinow, *Bobby Fischer Goes to War*, 200.
42. Edmonds and Eidinow, *Bobby Fischer Goes to War*, 201.
43. *Bobby Fischer Against The World*, 55:39 to 55:46.

44. Edmonds and Eidinow, *Bobby Fischer Goes to War*, 205.
45. Edmonds and Eidinow, *Bobby Fischer Goes to War*, 223-224.
46. Edmonds and Eidinow, *Bobby Fischer Goes to War*, 287.
47. Bobby Fischer, interview by Johnny Carson, *The Tonight Show With Johnny Carson,* NBC, November 8, 1972.
48. "Interview with Bobby Fischer," 1972.

Part III: Pop Culture

In the third and final section of the book, our authors explore transformations in production and representation practices in American television, and how those both reflected and shaped broader cultural trends in American society. These chapters demonstrate how television producers embraced new genres that both expanded and delimited the representation of American society. They highlight the importance of not just programming, but the changing industrial conditions of television production defined what Americans found on their television screens.

From the perspective of arts and entertainment, the prevailing idea seemed to be that television was an inferior medium, not designed to create art. Instead, critics hailed television as 'chewing gum for the eyes,' an enjoyable, but ultimately hollow experience. Even within television, many producers and executives considered certain types of content 'beneath them,' such as children's shows, or soap operas, both designed for markets that did not typically control family finances—children and housewives. The more respectable option was programming that catered to the sensibilities of what was considered the primary market: middle-class adult men.

America Reconciles with Vietnam Veterans: Changing Portrayals of Veterans in American Television Shows Under Reagan

JOSHUA BUCKINGHAM

"When I created Magnum, I got thousands of letters from Vietnam veterans thanking me
for portraying Vietnam veterans as something other than killers, and drug addicts, and crazy, and unable to function in society"

– Donald Bellisario, Co-Creator of *Magnum P.I.*[1]

American primetime television in the 1980s experienced a boom in action programming on the screen. Newly emerging shows such as *Magnum P.I.*, *The A-Team*, and *Miami Vice* shared a common trait: they all starred characters that put American military veterans at center stage. These featured "characters of redemption," who, due to the trauma they experienced during their time in service or from experiences in the civilian world or both, sought to do the right thing, help others, and return to a life of (somewhat) normalcy. These stories sought to dramatize the lives and stories of actual Vietnam veterans in the aftermath of the war. Such television shows not only represented a shift in positive public reception towards the military and its veterans; in some ways, they even produced this shift and facilitated the normalizing of relations between veterans and the public.

This chapter traces the representation of veterans from the post-World War II era to the 1980s. First, I examine veteran portrayal in television shows in the "golden age" of television in the 1940s and 1950s, in shows like *Crusade in Europe* and *Leave it to Beaver.* These shows appeared in the aftermath of World War II and were some of the first depictions of veterans on the television screen. I then turn to television shows produced during the Vietnam War, most notably *M*A*S*H.* While certainly not the only show produced during the war, *M*A*S*H* was a behemoth in terms of reception and popularity among viewers; it also focused specifically on the experience of soldiers only slightly removed from combat, and for many commentators reflected the anti-war sentiment many Americans at the time felt. Finally, I examine shows created in the aftermath of that war, including *Magnum P.I.* and *The A-Team*, among others. During this period, television programming both reflected and shaped American society and culture, mediating the escape from "Vietnam Syndrome" in the final stages of the Cold War. Not only does this give context on conflicts in which the United States was involved during and after WWII, but more important, it explores how the representation of these conflicts influenced public sentiment towards the military, as well as how such sentiments influenced the portrayal of American veterans on prime-time television during the Cold War.

WORLD WAR II AND THE AMERICAN HOMEFRONT

After the turbulence of the 1920s and the hardship of the Great Depression, many American experienced World War II (WWII) as a unifying experience. Americans contributed towards the war effort in a number of ways, through work in the defense industry, investment in war bonds, or taking belt-tightening measures such as growing victory gardens or recycling scrap metal.[2] There were also those who served in the armed forces: the U.S. Census Bureau estimated that close to sixteen million men and women served in the armed forces during WWII. Some voluntarily joined the military following the attack on Pearl Harbor and the intensification of the war in Europe, but sixty-one percent of the armed forces were called upon by the nation as draftees.[3] Initially, Americans did not uniformly commit themselves to the war. Most of the dissenters in the United States before 1941 included members of the Communist Party of the United States of America (CPUSA) and those who supported isolationist American policies.[4] Events in 1941, however—the invasion of the Soviet Union in June and the attack on Pearl Harbor in December—began to melt away anti-war sentiment in favor of a more hawkish view of American involvement in the war.[5]

By the end of the war, the conflict had become a "good war:" a fight for democracy against tyranny. The United States had responded to deliberate physical and ideological attacks, and the justice of its cause was confirmed by the virulence of the Nazi war of destruction, as well as emerging details about the Holocaust. Scenes of parades in American cities honoring WWII veterans and celebrating victories in Europe and Asia, were commonplace immediately after the war. Their contributions to American society were so great that the generation that fought in WWII would be dubbed the "Greatest Generation," a term popularized by the American broadcaster Tom Brokaw.[6] Just twenty years later, however, political conditions and American perceptions towards the military had changed drastically. Cold War conflict, including the stalemate on the Korean peninsula during the Korean War and the constant state of terror faced by citizens due to fear of nuclear war, had decreased American morale. Then, American involvement in the Vietnam War further threatened American perceptions of the military. American forces were mired in a brutal conflict halfway across the world, the purpose of which seemed to have very little to do with the interests of the everyday American, sparking an anti-war movement among civilians, students, activists, and even veteran dissenters, many of which had served in Vietnam. By 1970, American sentiment against the war had reached an all-time high. Veterans, many of whom had been conscripted into service during the Vietnam War, found themselves in a two-front conflict. One was in the jungles of Vietnam, while the other was on the homefront. Subject to the brutal conditions and horrors of combat, as well as the hazy moral choices of the American military effort, veterans believed that their fellow countrymen had turned their backs on them. While their WWII counterparts were lauded in the streets and treated as heroes, Vietnam veterans were subject to unfair scrutiny regarding their involvement in a war that they had little control over.

THE VIETNAM WAR: OVERVIEW AND CULTURAL IMPACT

The Vietnam War was an example of postwar decolonization caught up in the global Cold War rivalry. In the aftermath of the two World Wars, the declining strength of European states and rising movements for independence led to the collapse of empires across the globe. Vietnam was no exception, and in 1946 French colonial forces became entrenched against Viet Minh troops barely a year after victory in Europe in the First Indochina War. From 1946 to 1954, the French fought to regain control over its lost colonial territories but were ultimately defeated by the Viet Minh and forced to pull out of Vietnam in embarrassment.[7] The 1954 Geneva Convention established a line on the 17th parallel, temporarily dividing North and South Vietnam until national elections could be held to reunify the nation in 1956. Those elections were cancelled by the southern Diem regime with US support. In turn, the Viet Cong, a communist guerrilla group with the backing of North Vietnam, launched an insurgency against the South Vietnamese government. This started the Second Indochina War, later known by Americans as the Vietnam War.[8]

Under Presidents Eisenhower and Kennedy, American interest in Southeast Asia grew, with the primary policy being to reduce communist control and influence over countries in the region. The Truman-era policy known as "containment" had justified American involvement in the Korean War (1950-53) and spurred an increased interest in foreign political affairs.[9] Though the United States represented itself as an anti-imperial country, the American containment doctrine mandated that the US meet communism wherever it emerged, leading to paradoxical engagements in world conflicts.

During the earlier years of the war, the US involvement in the conflict was minimal, consisting of providing military advisors and arming and training the South Vietnamese for the anti-communist Army of the Republic of Vietnam (ARVN). In 1962, the United States established the Military Assistance Command, Vietnam (MAC-V), fulfilling a promise to increase American involvement in the conflict by establishing an official military command within the country.[10] In 1964, American support for the war in Vietnam solidified after the Tonkin Gulf Incident (an incident that would prove to have had some parts of it fabricated) when three North Vietnamese torpedo boats allegedly attacked the USS Maddox.[11] This incident led quickly to the enacting of the Tonkin Gulf Resolution (which had been drafted months prior to the incident) on August 10, 1964, officially authorizing US combat operations and troop deployments to South Vietnam and the bombardment of North Vietnam.[12] In 1965, the first Marines landed in Da Nang, igniting direct combat between the U.S. military and communist forces.[13] At this time, most Americans did not know the history of the conflict in Vietnam and were unlikely to be able to point out the country on a map. American sentiment towards the war was generally neutral in the earlier years from 1965 to 1967, but, the United States deepening commitment in the war, the persistence of the military draft, and American combat losses sparked debate, protests, and ultimately the emergence of an antiwar movement.

The Vietnam War is often referred to as the "first televised war," with the media coverage of the conflict posing a challenge to the American military and policymakers. This was due to the fact that Americans on the homefront were now seeing images from the combat zone on a scale never seen before, often before the American government had time to make its own assessments.[14] These images, as well as the nightly listing of Americans Killed in Action (KIA), Wounded in Action (WIA), and Missing in Action/ Prisoners of War (MIA/POW) on television news programs allowed for public dissent towards the war to grow. A surprise offensive by the North Vietnamese during the holiday of Tet in 1968—the Tet Offensive—represented a turning point in the war, catching Americans and

South Vietnamese alike off guard. It also was an embarrassment for the US, with communist saboteurs managing to get inside the American embassy in Saigon. Even though the mission was suicidal in nature, the message was clear: the US would not win against a determined guerilla opposition that could slip in and out of enemy territory with ease. Walter Cronkite, a CBS news anchor known as the "most trusted man in America," famously predicted that "the bloody experience of Vietnam is to end in a stalemate."[15] Cronkite's assessment mirrored American frustration with the war, fearing that American foreign policies had entrapped Americans into fighting a prolonged war with no clear sign of victory.[16] President Lyndon B. Johnson saw the writing on the wall and lamented, "If I've lost Cronkite, I've lost middle America."[17] This realization was an important factor in his decision not to run for reelection in 1968, giving rise to Richard Nixon's presidency.

Despite the rising tide of anti-war sentiment, in 1970 President Nixon announced an expansion of the war that further fueled anti-war demonstrations in the US and abroad. This announcement followed the year-long covert Operation Menu, which saw the heavy bombing of Cambodia by the United States Strategic Air Command (SAC).[18] The US military began to operate in desparate ways: conducting an incursion into Cambodia, determined to root out Viet Cong operating there with the conviction that combat operations in Cambodia would help to bring an end to the war in Southeast Asia.

Antiwar demonstrations reached their peak in 1970, marking the beginning of the end of the American war effort in Vietnam. On May 4, a demonstration against the Vietnam War at Kent State turned bloody. Ohio National Guardsmen, launching tear gas toward the crowds of demonstrators and moving in formation, killed four students and injured nine.[19] Media coverage of this event sent Americans across the US into a fury, ramping up anti-war demonstrations even more. The incident, which involved American military personnel firing upon American civilians, came to haunt American domestic policy and helped to convince the president to withdraw US combat forces. On May 8, 1972, President Nixon signed the Paris Peace Accords and agreed to withdraw US troops from Vietnam within the year through a process called "Vietnamization."[20] Ten months later, on March 29th, 1973, the last American troops withdrew from Vietnam, leaving the South Vietnamese to carry out the remainder of the conflict in the hope that a non-communist government in Vietnam would survive.[21] Amid the collapse of ARVN forces, the North Vietnamese Army captured the capital of South Vietnam, Saigon, on April 30, 1975.[22]

Veterans of the Vietnam War did not receive the same warm welcome that WWII veterans had enjoyed upon their return from Europe and Asia after the war. Instead, confusion and anger over the goals and means of the military conduct of the war led to a different response to returning veterans. Some were met with little attention, while others claimed to be confronted by their fellow countrymen. The sense of loss in Vietnam was so extreme that it constituted a term known as "Vietnam Syndrome," which describes the hesitancy of American policymakers to engage in foreign conflicts to avoid embarrassments similar to Vietnam. This should not be mistaken as Vietnam Stress Syndrome, a term that described combat trauma. After coming home, many Vietnam veterans experienced psychiatric symptoms that later became known as Post-Traumatic Stress Disorder (PTSD). Undiagnosed trauma and depression became compounded by the isolation many felt upon their return due to the unpopularity of the Vietnam War. Although the Carter administration provided more resources and support for veterans, Ronald Reagan made the plight of Vietnam veterans central to his election campaign. During the 1980s, a cultural shift occurred in the United States, placing veterans back in a positive light and belatedly welcoming them home.

AMERICAN VETERANS ON TELEVISION BEFORE VIETNAM

Television was a relatively new entertainment source for the average American family in the postwar era, with television sets becoming a staple in homes starting in the late 1940s and early 1950s.[23] Five million American families had already purchased a television set by 1950, primarily due to the technology becoming more available and affordable to the nuclear family. In the earlier days of television in the United States, there were just four networks for viewers to watch television: the American Broadcasting Company (ABC), the Columbia Broadcasting System (CBS), the National Broadcasting Company (NBC), and the DuMont Television Network.[24] These networks aired shows like *The Texaco Star Theatre* (NBC), *The Ed Sullivan Show* (CBS), and *The Lone Ranger* (ABC) that were targeted at a family audience.[25]

While early programming was often fictional and meant for entertainment for the family, Americans were also not spared representations of warfare on the television screen. Through documentary-style shows, like *Crusade in the Pacific* (ABC) and *Victory at Sea* (NBC), Americans could learn more about WWII and the battles that were involved from the comfort of their homes. *Crusade in Europe* (ABC), a television adaptation of then-General Eisenhower's book of the same name, used archival footage of World War II to tell the story of how the war started and American involvement in crushing the Nazi war machine. In it, American G.I.s are portrayed as tough and determined to defend democracy, even at the expense of their own lives, which deeply resonated with the millions of American families who had lost loved ones in the war.[26] All of these documentary-style shows built a narrative consensus around the conduct and purpose of the war: the war was built on American resolve to expose and expunge the evil of the Axis powers. They did not shy away from showing American soldiers in their most vulnerable state, highlighting their suffering and the horrors of combat that soldiers experienced on the battlefield. Above all, these shows interpreted the war through the lens of American victory, something that would be exceptionally harder to do with the war in Vietnam.

On network television entertainment shows, depictions of American veterans drew upon popular narratives as seen in the war documentaries. Shows depicted veterans as pillars of their communities. Beloved television characters like Ward Cleaver (Hugh Beaumont), father of the eponymous Beaver (Jerry Mathers)from *Leave it to Beaver*, was role model for his children, whom he taught to abide by the law and be respectful towards others. Ward was a white-collar worker, middle class, and happy with his family life. He was a modest man, and like many veterans of the time, rarely spoke about his experience in WWII as a Navy Seabee. His experience and lessons learned in the military, however, did account for the authority he had within his community and household, especially over his children. Ward embodied the trope of American masculinity at the time, serving as the go-to father, family man, and good citizen; something that other sitcom-style television shows imitated and made their own during that era. While not all veterans returning from WWII shared Ward's postwar experience, the creators of *Leave it to Beaver* likely felt that these demographics would have fit the postwar consensus on middle-class society, making Ward more relatable to the viewer. Ultimately, Ward Cleaver's depiction in *Leave it to Beaver* drew upon and reinforced positive public perceptions of veterans in post-WWII America. In *Leave It to Beaver*, as well as other sitcoms of the day, like *The Andy Griffith Show*, the idea of "raising the next generation" is prevalent in many of the episodes. Both shows focus on a father-son dynamic and use this dynamic to speak about many important life lessons to help the next generation grow into good members of society. Both fathers are constantly challenged by their children to address questions like telling the truth and treating others like they would like to be treated but rise to the challenge and do so with a positive attitude.

Andy Griffith's portrayal of Sheriff Andy Taylor in *The Andy Griffith Show* is another depiction of an American WWII veteran in a role that has positive influence and authority over others in a community. Throughout the series, Sheriff Taylor and his deputy/cousin, Barney Fife (Don Knotts), encounter mainly comedic situations within their community, and the plot revolves around their efforts to deal with them. Perhaps the greatest contribution of Sheriff Taylor is through the lessons he teaches his son, Opie (Ron Howard), and in turn the audience. This role as a supportive dad, similar to Ward Cleaver's role, speaks to the father figure trope that was common in earlier sitcoms. Sheriff Taylor's military service is not key to the plots of the series, although it is mentioned in a few episodes.[27] What is important about this, however, is that Andy Taylor's veteran experience underpins his character: similar to Ward Cleaver in *Leave it to Beaver*, Sheriff Taylor could be characterized as a modest man, only bringing up his time in the armed forces when joking around with other characters. He enforces the law in his small town, serving as sheriff and Justice of the Peace, yet remains approachable to his neighbors, who feel comfortable enlisting him in their times of need. Taylor held a sense of responsibility and authority over the town he and his family lived in. Like Ward Cleaver, Sheriff Andy Taylor taught his family to uphold the values of integrity, honesty, and hard work. However, Sheriff Taylor took it a step further than Ward by also instilling the value of commitment to service into his son and those around him. Ward Cleaver is the "man of the house", willing to teach his children life lessons; Andy Taylor does this for his son, Opie, as well, but also actively teaches the importance of service to the community through his prominent role as Sheriff. To iterate, a common concept for both of these shows is the father figure trope, all the more authoritative and authentic due to their backgrounds as veterans. They serve as surrogate fathers to the nation: teaching the next generation to be good citizens.

Another character type emerged in shows that focused not on veterans, per se, but rather soldiers serving in military roles, often played by veteran comedic actors. The *Phil Silvers Show*, sometimes referred to simply as Sergeant Bilko (after the main character), was an incredibly successful and popular sitcom with the American public, winning three Emmy Awards during its run.[28] The show starred Phil Silvers as Master Sergeant (MSgt.) Ernest Bilko, United States Army, and unlike the shows discussed earlier, MSgt. Bilko was not a model citizen but rather a con artist of sorts. He is best described "as a scheming moto pool sergeant fleecing his shambolic platoon . . . in the service of get-rich-quick schemes that never quite come off."[29] A product of post-Korean War society, *The Phil Silvers Show* sought to bring laughter and joy to the audience, perhaps sometimes seen as a distraction from the tumultuous world that perched outside the door of the family household during the Cold War. MSgt. Bilko and his posse were not typical veterans of the time; in fact, they were technically still in the military at the time of the show. That being said, the perception of this show towards veterans and the military could be classified more so in terms of anti-militarism. While other shows of the time contained messages of morality and family values, *The Phil Silvers Show* spends its time dealing in mischief and being above the law. While this may not necessarily be perceived as a negative portrayal of veterans at the time, it was a departure from the dignified treatment of Cleaver and Taylor as described above, highlighting an irreverence that would continue to grow into the 1960s.

McHale's Navy was a sitcom in the 1960s that took great inspiration from *The Phil Silvers Show*, even hiring the same actors and writers to maintain a similar comic atmosphere. The show followed Lieutenant Commander (LCDR) Quinton McHale (Ernest Borgnine), United States Navy, who maintained an attitude and actions similar to that of MSgt. Bilko throughout the duration of the series. Like Bilko, McHale and his crew were often too interested in partaking in mischievous activities to truly focus on the mission at hand. Unlike Sheriff Andy Taylor and Ward Cleaver, McHale and his crew were the opposite of the good, law-abiding American veteran. While series like *The Phil Silvers Show* and *McHale's Navy* was focused on the comedic aspects of military life, they strayed away from the narrative of the "good citizen" trope. While a stricter view on this portrayal may say it was a negative and/or harmful portrayal of veterans in society, a more light-hearted view could see this as American society saying that "boys will be boys." The camaraderie aspect of the military is one that is unlike any other profession, and perhaps these shows tried to imitate this brotherhood through comedic scenarios.

While *The Andy Griffith Show* and *Leave It to Beaver* were shows that kept family values in mind, shows like *The Phil Silvers Show* and *McHale's Navy* took a more comedic approach towards the American sitcom genre of the 1950s/60s. Whereas the main characters of the aforementioned shows were more focused on the father figure trope and service to the community, the main characters in *The Phil Silvers Show* and *McHale's Navy* sometimes held questionable morals. Moreover, while *The Andy Griffith Show* and *Leave It to Beaver* (shows that arguably were more about learning lessons and aimed at family values) were set in hometown communities, these other two shows were set in a primarily military setting and also spent more time on comedic deliveries than lessons. In this, the split on how veterans are portrayed in television at the time became more prominent, with one side aiming to have veterans be model citizens, while the other side used military situations more so as comedy pieces. The latter side would become more prominent over the years leading up to the Vietnam War. However, once dissatisfaction with the war spread near unanimously throughout the American homefront, a new portrayal of the American veteran would take center stage: one that was broken and dismayed.

VETERANS ON VIETNAM-ERA SHOWS

Gilligan's Island first aired on September 26, 1964, on CBS; just over a month after the Tonkin Gulf Resolution was enacted.[30] In this sitcom, there are two veterans starred as the main characters: Bob Denver portrayed Willie Gilligan and Alan Hale played Captain Jonas "Skipper" Grumby. Similar to *McHale's Navy* and *The Phil Silvers Show*, *Gilligan's Island* used these veterans for comic relief. Gilligan and Skipper served together in the US Navy, with Skipper being credited with saving Gilligan's life during their service. The heroics ended there, however, as these two characters stuck to physical comedy sketches throughout the duration of the series. Gilligan and Skipper represent the most absurd representation of veterans since WWII, but it did not survive the Vietnam war. While the series began just after the enactment of the Tonkin Gulf Resolution, it was canceled before the Tet Offensive escalated the war.[31] The war's impact on public opinion is perhaps best seen in the comparison of *Gilligan's Island* to the Larry Gelbart's *M*A*S*H*: both shows thematize military situations for comic relief, but offer very different underlying tones. While *Gilligan's Island* depicts carefree adventures, *M*A*S*H* digs deep into the anti-war sentiment felt at the time in America.

No television show created during the Vietnam War quite encapsulated America's ongoing resentment towards war quite like *M*A*S*H* did. Created in 1972, near the end of America's involvement in Vietnam,

*M*A*S*H* was a sitcom featuring an ensemble of characters including American doctors and military personnel deployed during the Korean War.[32] While the show was set during the Korean War, it unmistakably offers commentary on the war in Vietnam. The most obvious of this is the actual setting: a Mobile Army Surgical Hospital (MASH) unit where the main protagonists constantly are reminded of the ongoing war while attempting to save wounded soldiers. While the characters throughout the series react to comedic situations, engaging in hijinks around the MASH in between taking in casualties, some episodes dig deeper. The episode titled, "The Interview," is styled after documentary style wartime battlefield interviews conducted during the Korean and Vietnam Wars: the main characters spent the episode answering questions (often ad-libbed) ranging from the ongoing Korean conflict to what they miss most about home.[33] Perhaps the greatest aspect of this episode is its haunting, as the characters reflect on and recount the MASH experience only slightly removed from the front lines. In an interview over 40 years later, the actors described how much of the episode was ad-libbed, with the actors improvising answers in character.[34] This arguably made the reactions more human and the emotions rawer, creating tense scenes for the viewer. *M*A*S*H*'s success was in part due to its ability to mirror the sentiment of the American people during a time of turmoil, but over the next decade, veterans became a major focus of media, films, and public rituals.

VETERANS IN REAGAN'S AMERICA

The Reagan era was a period of rising conservatism in America, including renewed Cold War conflict, and domestic policies accompanied by an "America first" message. Reagan took a hardline approach to what he perceived as "Soviet aggression," reduced taxes for the wealthy, and initiated "Reaganomics," a program for growing the uses and access of technology in America. Most important to this chapter, however, was Reagan's stance towards veterans and the military. Reagan viewed "the Vietnam War as 'a noble cause' in which the United States had sought to defend 'a small country newly free from colonial rule . . . against a totalitarian neighbor bent on conquest.'"[35] Reagan felt strongly that the American people had unjustly isolated Vietnam veterans from society, and he sought to right this through the adoption of veteran-friendly policies like the expansion of the Veteran Affairs system and the completion of the Vietnam War Memorial (although plans to build the memorial began under Carter's presidency). It was during this time that a cultural shift to make amends with veterans of the war began; television helped mediate this shift.

The first positive depiction of Vietnam veterans in a television show is perhaps *Magnum P.I.*, starring Tom Selleck as former US Navy SEAL Thomas Magnum. Production on the show began in December 1980, just before Ronald Reagan's inauguration, during a time when masculinity in America was being redefined and action television series became a staple in American television culture. This time, military service was not assumed, but rather was a central, defining feature of the title character. Where Ward Cleaver and Sheriff Andy Taylor rarely spoke on their time in the service in the 1950s, Thomas Magnum's service in the US Navy as a SEAL and Naval Intelligence Officer is central to his mission. Throughout the series, multiple episodes cut back to scenes of Magnum in the jungles of Vietnam, fighting alongside other members of his SEAL team, often losing someone important to him; these are experiences that haunt his character in the present day. But, in contrast to contemporary cinematic visions of the unstable Vietnam veteran incapable of living a civilian life—such as John Rambo (Sylvester Stallone) in *First Blood*—Magnum is a capable and reliable neighbor, friend, and private investigator. Though affected by his experiences in Vietnam, he is still able to function within society; Rambo cannot. It is this sense of normalcy that made *Magnum P.I.* appeal to veterans of the Vietnam War: they could relate to Magnum's portrayal of the veteran and were no longer demonized on the screen.

Similarly, *The A-Team* portrayed four Army Green Berets helping civilians take on crime in the Los Angeles area.[36] Perhaps one of the most action-driven television shows of the 1980s, the four characters often engage with antagonists in a "blaze of glory," utilizing machine guns and explosives often. The background of these characters is important to understand their motives behind helping others throughout the show. All four members of "The A-Team" had served with one another during the Vietnam War, but due to a mission gone wrong, the team had to go underground. They were on the run from the federal government for crimes they did not commit. Their members included Colonel (COL) John "Hannibal" Smith (George Peppard), Captain (CPT) H.M. "Howling Mad" Murdock (Dwight Schulz), First Lieutenant (1LT) Templeton "Faceman" Peck (Dirk Benedict), and Sergeant First Class (SFC) Bosco "Bad Attitude" Baracus (Mr. T.) Each of these characters had a specific skill set; only together could they accomplish the mission. The Vietnam War is a defining facet of *The A-Team* and how the show's characters interact with one another and their environment. All of the members of the team were impacted by the war, experiencing the loss of comrades and the "betrayal" of their own military against them. In the show's epsiodes, they seek redemption and to help those who need it most. Their portrayal as veterans is sometimes stereotypical, specifically during action sequences. However, what *The A-Team* captures about veterans is their unbreakable bond and comaraderie. Throughout the series, the team faces challenges that if done by just one of them, are not enough. Together, though, the team is up for the task at hand. While sometimes cartoonish in style, *The A-Team* is another example of the post-Vietnam revitalization of veterans in television series.

CONCLUSION

The Vietnam War represents a dark period in American history, not only for the failure of the combat operations during the conflict but also for the way veterans were scrutinized for their involvement by their fellow Americans. When Vietnam veterans returned home from the battlefield, they were not greeted with the same parades and celebrations their WWII predecessors had enjoyed. Instead, many felt excluded by the public and ashamed for their participation, despite having been conscripted, leading many to feel isolated and disconnected from post-Vietnam War society. There is a clear shift in the portrayal of American military veterans in television series depending on the time period and the cultural atmosphere towards the military. From the post-WWII father trope and/or model citizen, to the switch to a more comic or antiwar portrayal that contributed towards the downward trend of respect for the veteran during the Vietnam War, to finally the emergence of the veteran as the hero in post-Vietnam America. While some creators of television series may not consider how their characters may impact the groups of people they portray, creators of shows like *Magnum P.I.* understood this impact, making Magnum's character one of redemption and reintegration with society. The portrayal of the veteran in television during the Cold War is just one example of how television can impact public perception, and vice versa, on certain groups of people and topics.

Notes

1. FoundationINTERVIEWS, "Donald Bellisario discusses the portrayal of Vietnam vets on "Magnum, P.I." EMMYTVLEGENDS.ORG," *YouTube Video,* 4:27. July 17, 2015.
2. Jonathan E. Vespa, "Those Who Served: America's Veterans From World War II to the War on Terror," in *This Is Who We Were: In The 2010s*, edited by Grey House Publishing (Amenia, NY: Grey House Publishing, 2020), 379; Phyllis Raybin Emert, *World War II: On the Homefront* (Carlisle, MA: Discovery Education, 1996), 27.
3. "US Military by the Numbers," Research Starters, National World War II Museum New Orleans, https://www.nationalww2museum.org/students-teachers/student-resources/research-starters/research-starters-us-military-numbers.
4. Maurice Isserman, "War in Europe, August 22 to December 1939," in *Which Side Were You On? The American Communist Party During the Second World War* (Middletown, Conn.: Wesleyan University Press, 1982), 33.
5. Isserman, *Which Side Were You On,* 102.
6. Tom Brokaw, *The Greatest Generation* (New York: Random House, 1998).
7. Ronald B Frankum, "Vietnam during the rule of Ngo Dinh Diem, 1954-63," in *The War That Never Ends: New Perspectives on the Vietnam War*, ed. David L. Anderson and John Ernst (University Press of Kentucky, 2007), 128.
8. Frankum, "Vietnam," 134.
9. George C. Herring, "America and Vietnam: The Unending War," *Foreign Affairs,* Vol.70, no. 5 (1991): 107.
10. Graham A. Cosmas, "A Deepening Commitment and a New Command," in *MACV: The Joint Command in the Years of Escalation, 1962-1967*, (Washington, DC: Center of Military History United States Army, 2006), 27.
11. Adam Roberts, "The Fog of Crisis: The 1964 Tonkin Gulf Incidents," *The World Today*, Vol. 26, No. 5 (1970): 211.
12. Roberts, "Fog of Crisis," 209.
13. Robin Gentry and Phil Martin, "Marine Corps Engineers in Vietnam," in *The Military Engineer*, Vol. 108, no. 703 (2016): 28.
14. Michael Mandelbaum, "Vietnam: The Television War," *Daedalus*, Vol. 111, No. 4 (1982): 157.
15. CBS Evening News, "50 years ago: Walter Cronkite calls for the U.S. to get out of Vietnam," *YouTube Video*, 0:54, February 28, 2018.
16. CBS, "Walter Cronkite calls for the U.S. to get out of Vietnam."
17. William M. Hammond, "The Tet Offensive and the News Media: Some Thoughts on theWilliam M. Hammond, "The Tet Offensive and the News Media: Some Thoughts on the Effects of News Reporting," *Army History,* no.

70 (2009): 11. http://www.jstor.org/stable/26373973.

18. (working paper, History of Advertising Archives, Faculty of Commerce, University of British Columbia, Vancouver, 1980).William P. Head,"Menu Bombing," in *War From Above the Clouds: B-52 Operations during the Second Indochina War and the Effects of the Air War on Theory and Doctrine*, (Fairchild Paper, Air University Press, Maxwell Air Force Base, Alabama, 2002).

19. Christopher J. Broadhurst, "'We Didn't Fire a Shot, We Didn't Burn a Building:' The Student Reaction at North Carolina State University to the Kent State Shootings, May 1970," *The North Carolina Historical Review*, Vol. 87, No. 3, (2010): 288.

20. Chester Pach "'Our Worst Enemy Seems to Be the Press': TV News, the Nixon Administration, and U.S. Troop Withdrawal from Vietnam, 1969-1973," *Diplomatic History*, Vol. 34, No. 3 (2010): 565.

21. Ibid.

22. Sucheng Chan, ed., *The Vietnamese American 1.5 Generation: Stories of War, Revolution, Flight and New Beginnings* (Philadelphia: Temple University Press, 2006), 57.

23. Albert Abramson, *The History of Television, 1942 to 2000*, (Jefferson, NC: McFarland & Company, 2003), 18.

24. Ibid, 32.

25. James L. Baughman, "Mr. Spectacular," in *Same Time, Same Station: Creating American Television, 1948-1961*, (Baltimore: The John Hopkins University Press, 2007), 82-86.

26. *Crusade in Europe*, directed by George Black (The March of Time, 1949).

27. In episode 4 of season 1, "Ellie Comes to Town", Andy mentions that he was over in France during WWII while looking at perfume made in Paris. In episode 9 of the same season, "A Feud is a Feud", Andy again says that he served in France during WWII after mentioning he picked up some basic French during his time deployed there.

28. n.a., "The Phil Silvers Show," Internet Movie Database, last accessed 13 June 2024, https://www.imdb.com/title/tt0047763/

29. John Mundy and Glyn White, "Television Comedy" in *Laughing Matters: Understanding Film, Television and Radio Comedy*, (Manchester: Manchester University Press, 2012), 112.

30. Laura Morowitz, "From Gauguin to Gilligan's Island," *Journal of Popular Film and Television*, Vol. 26, no. 1 (1998): 2–10.

31. Walter Metz, "Introduction: A Minimalist Allegory," *Gilligan's Island* (Detroit, MI: Wayne State University Press, 2012), 1.

32. James H. Wittebols, "Episode Guide," in *Watching M*A*S*H, Watching America: A Social History of the 1972-1983 Television Series*, (Jefferson, NC: McFarland & Company, Inc., Publishers, 2003), 161.

33. "The Interview," directed by Larry Gelbart, *M*A*S*H*, season 4, episode 24,

20th Century Fox Television, 1976.

34. MeTV, "M*A*S*H Episode Deep Dive | "The Interview," *YouTube Video*, 3:11, July 2, 2020.

35. Lou Cannon, "Reagan: 'Peace Through Strength'," *The Washington Post*, August 18, 1980.

36. TalkerOne, "A-Team Intro High Quality," *YouTube Video*, 1:33, February 10, 2009.

Who Has The Power?: The Unique Landscape of Children's Television in the 1980s

AAIDIN FINEFIELD

On November 10, 1969, the new production company Children's Television Workshop (CTW) celebrated the premiere of its first program: *Sesame Street*. The educational show was highly anticipated by the press and critics alike, in the hopes that it would be a game-changer for children's television, due to their disappointment in the effectiveness of educational television at the time.[1] The show utilized a multimedia format, featuring live action sequences (featuring humans and puppets) intercut with animated segments. The goal was to educate children both about topics such as the alphabet or counting, which were fairly common on other children's shows, but also more advanced concepts that were practically unrepresented onscreen like death or disability, while still being entertaining and memorable.[2] Additionally, *Sesame Street* featured a diverse cast of people and was thoughtful about not reproducing cultural or ethnic stereotypes. Alongside producers, experts on child development, education, social issues, and more, all had an "equal status in the decision-making process" of the show. Along with *Mr. Rogers' Neighborhood*, this proved a rare exception to the norm of producers being the only decision-makers in television.[3] Determined to improve children's programming, *Sesame Street* remained open to constructive criticism. For example, Hispanic activists accused creators of subpar bilingual representation in their early years, production staff at *Sesame Street* worked to include more Hispanic people both on-screen and behind the scenes.[4]

Sesame Street thus exemplified a new approach to children's television programming. Joan Ganz Cooney, one of the founders of the CTW, was convinced that "educators were virtually ignoring the intellect of preschool children" by prioritizing protecting kids from possibly challenging or controversial topics rather than teaching them skills.[5] As a result, television had earned the reputation of creating passive kids, not challenging them to take initiative; many contemporaries even believed that "passively watched television could never teach."[6] However, *Sesame Street* directly challenged and defeated this notion. Kids who watched *Sesame Street* and *Mister Rogers' Neighborhood* as preschoolers had an average 0.25 higher grade than their peers in high school and came to school much more excited to learn.[7] Moreover, it was widely popular: by 1976, an average of nine million children watched *Sesame Street* every day.[8]

The show was popular among critics, parents, and children, and was designed to fulfill a noble goal—the education of America's children—yet it remained an outlier in the history of children's television. By the 1980s, the majority of programming was less educational, lower quality, and more consumerist than *Sesame Street.* Advertisers, not educational professionals, and profits, not intellectual outcomes, shaped this programming. Nor did this situation improve over time. *Sesame Street* remained an outlier in, not a trendsetter for, the children's television industry. In the 1980s, the Reagan administration deregulated television, and producers and advertisers changed the business of children's television in pursuit of greater profits, which in turn changed the landscape of children's television itself.[9] The main driver for this new direction in the industry was risk management, a process that essentially sought to manipulate the market to maximize profits. Understanding the role of risk management also offers a new way of understanding the way children's television changed the way it presented itself as educational and handled race and gender.

A BRIEF HISTORY OF TELEVISION

As mass adoption of television swept across the nation in the 1950s, more Americans became familiar with this revolutionary new technology. Producers experimented with forms and content and the broadcast day expanded. The amount of content available on television began to balloon, including the amount of advertising. In 1941, the very first commercial played on television after a baseball game; the rise of network television—and the end of the sponsorship era—allowed more companies to "sponsor" or advertise on television.[10] As television grew up, it became expensive, but it also offered great rewards in terms of profits. As profits rose, greater scrutiny was placed on the industry.

Television marketed towards children became an area of increased scrutiny. Criticism began loosely in the 1950s but rose to a much higher pitch by the 1970s and 1980s. Initial reactions to television in the 1950s were extreme. Many saw it with anxiety and panic while others heralded television as having great potential for education/entertainment.[11] Both sides had similar concerns though, and these criticisms usually revolved around some of the most contentious topics throughout American history that were being challenged significantly during the mid-to-late twentieth century. These were topics like race, sex, and consumerism along with how television programming instilled ideas about each topic.

At first, one of the major goals of children's television was education. Whether the programs *actually* were educational is a different issue, but the goal of the programs was education, and they were heavily on the rise. By 1967, there were 140 non-commercial educational television stations (not all were built for home viewing; instead a significant number were broadcast into schools, colleges, and industrial workplaces).[12] Educational television for children included examples such as *TV Kindergarten* and *Merlin the Magician,* which taught children reading, numbers, and slightly more abstract concepts like cooperation.[13] Despite the fact that children's television had an educational focus during this time period, many critics still saw it as wasteful and complained that it "takes of a viewer's time without giving anything in return."[14] so it is important to remember that quality was slowly getting better while being mostly panned by the adult population.

The 1960s and 1970s were also a period when many hallmarks of network television emerged. American networks such as the American Broadcasting Company (ABC), the Columbia Broadcasting System (CBS), and the National Broadcasting Company (NBC) were firmly established as major players in the television world by the mid-1970s; the Public Broadcasting Service (PBS) was launched in 1967; and color television arrived in the 1960s as well.[15] These developments were in reaction to the increased demand for television as companies rushed into this rapidly developing industry. With so much room to grow, there was much profit to be made. The paradox between consumerist growth and educational content was held at bay only by government regulations.

By the late 1960s, government officials began to take a greater interest in the significance of television in children's lives. In 1968, President Lyndon B. Johnson created the National Commission on the Causes and Prevention of Violence. This committee's mandate was to investigate the causes of societal violence, one of the subjects of interest being mass media.[16] The government's concerns of the government revolved mainly around depictions of violence and the way products were sold to children. The committee, as well as broadcasting's governing body, the Federal Communications Commission (FCC), encouraged broadcasters to self-regulate their programming. Broadcasters generally stayed away from depicting violence in a good or neutral light or basing programs around specific products for fear of what might happen if they pushed the envelope.[17] The quality of children's television slowly increased over the period of the 1970s due in part to the oversight of the commissions, and otherwise sought to remain uncontroversial: shows were produced, ran their time on television for a few years, and retired without challenging the landscape. That was the case until President Ronald Reagan's administration reshaped the regulatory environment for television.[18]

For example, one pressure group particularly focused on children's television successfully lobbied the FCC in 1971 to "closely examine commercial activities in programs designed for children."[19] The Action for Children's Television (ACT) was responding to violence and commercialism and advocating for the complete removal of violence and advertisements in children's television. But within a decade, their efforts were undone by a new administration determined to "de-regulate" American industries. Under the Reagan administration, the FCC reversed or lifted many of its rules.[20] By 1984, for example, the FCC had lifted regulations against completely unrestricted commercial airtime across the television schedule. This opened the door for studios to fully embrace inexpensive program-length commercials with debatable educational and aesthetic quality directed at children.[21] The ACT reported that their efforts had a sudden and negative downturn in success, and their advocacy now largely fell on deaf ears.[22]

The 1980s are important to the history of children's television due to its anomalous status, at least in comparison to the previous couple of decades, in the development of the medium. It was an anomaly because it moved in a significantly different direction in terms of corporate makeup, political goals, and marketing strategies. The changing rules of television held important consequences for the representation of consumerism, racism, and sexism, as will be discussed below.

THE ADVERTISERS' CONUNDRUM

Advertising has been central to American television since the 1940s. Before 1941, television networks were banned from running commercials, but that did not stop them from reading messages from sponsors while on air in 1939. Later, regulations allowed networks to sell blocks of commercial time, with revenues determined on a sliding scale based on the size of the likely audience. As an example, in 2016, national TV commercials cost around eight thousand USD on average, but ads during the Super Bowl cost around five million USD per thirty seconds.[23] This held true in children's television as well: advertisers were limited to commercial breaks to try and sell toys, clothes, food, candy, and more to kids—who were only attracted to shows specifically geared towards them.

By the 1980s, the expectations of the network era had begun to erode. Advancements in television technologies gave viewers more control over their own attention, and advertisers became nervous about the effectiveness of commercials for children. For example, the introduction of the television remote allowed the viewer to avoid or ignore commercial content, whether by "channel-surfing" or making use of the "mute" button—once a show cut to commercial, viewers could mute the sound of the TV and more easily ignore ads. In a similar vein, advertisers lamented the invention of the videocassette recorder (VCR).[24] Before the VCR, there was little a viewer could do to control their television experience, except turn the set on or off; the sound up or down; or change the channel. People were forced to sit through the exact same number of commercials as everyone else, but with the VCR you could record a show as it broadcasted, then fast forward through the commercials. Predictably, the VCR was hugely popular, and it had been adopted by fifty percent of US households by 1987.[25] Finally, the introduction of cable television and the multi-channel universe was also a huge hit to advertisers. Cable channels such as MTV, The Weather Channel, and C-SPAN challenged the monopoly of attention network broadcasters had on television viewers.[26]

In addition to innovations in television technology, market research also reshaped advertisers' expectations of television. Research helped advertisers re-conceptualize the role of the consumer. Conventionally, advertising had been directed at adults – the breadwinners and household managers of the family. But now advertisers conceptualized the parent or guardian as essentially a middleman for the actual consumer of the product, the child. Children's television allowed companies to fully realize the economic role of childhood, rather than it only being a social/cultural concept.[27] This worked in practice too: by 1994, a full decade after companies were allowed to run rampant on children's television, a group of surveyed parents stated that seventy-four percent of them had bought toys for their children based purely on the fact their child had seen a commercial for that toy.[28]

Yet market research also demonstrated that the current model for advertising was losing its efficacy. Some studies began to show that, by the age of seven or eight, kids largely began to ignore advertisements.[29] Before that age, children were unable to take in commercials critically and instead believed what the commercials were saying as "truthful, accurate, and unbiased."[30] By the time kids became school-age, they could tell the difference and likely began to resent the commercials for interrupting their shows, thus ignoring the commercials. In a world where viewers could isolate or ignore commercials' content through technological means or their own sense of disdain, advertisers realized that they needed to change the way they approached television.

To address the problem of commercials quickly being phased out of many kids' experiences with television, advertisers shifted their attention towards the programs that brought young viewers to the screen. *Sesame Street* (along with other shows being released around the same time) inspired advertisers, in a way: the characters within the show were incredibly popular, and merchandise based on them—toy versions of Bert & Ernie, Big Bird, Oscar the Grouch, and many more—was fairly instrumental in funding the non-profit production company that made *Sesame Street*.[31] The toy industry—the most important advertiser on children's television—quickly took note.

The toy industry is one of the oldest manufacturing industries in the United States, but it has always been incredibly unstable.[32] Traditionally, parents only bought toys during the winter holiday season and for birthdays. Because of this, the toy industry was forced to be almost purely seasonal. Moreover, as consumer expenditures in the winter holiday season grew, so did companies' efforts to meet demand. If, however, a company's toys for the season were not as popular as expected they were at serious risk of losing a lot of money.[33] By contrast, *Sesame Street* merchandise sold year-round. Toy companies then sought ways to produce year-round products too. Television programming became central to their efforts: suddenly, window shopping did not require the customer to leave the house, because the window always lived in their living room.[34] Television promised to sate the industry's strong interest in consistent sales, which improved their bottom line with much less risk. The main driving force for advertiser's interest in television programming was purely monetary.

CHARACTER LICENSING AND ITS CONSEQUENCES

One of the major effects of advertisers moving into broadcasting was a renewed focus on the practice of character licensing. Character licensing is an economic strategy that describes creators selling the right to use a character's image on any products to various companies. Character licensing was not new—the first licensed character was a stuffed animal based on *The Tale of Peter Rabbit* produced in in 1903—but it was significantly ramped up in the 1980s.[35] Licensed characters could appear on toothpaste, lunchboxes, clothing, stickers, candy, and much more. In conjunction with the rise of popular branded television channels directed towards niche audiences (see, for example, Nickelodeon, founded in 1979), the brands of children's television quickly became household names.[36] Beginning in the 1980s, shows such as *The Care Bears, GI Joe, The Smurfs,* and *Transformers* were created by advertisers, license groups, and broadcasters and licensed specifically to sell pre-existing products or to create a body of characters to bring to market.[37]

The very first licensed character franchise (not just one-off character licensing like *Peter Rabbit*), launched in 1980, and it was hugely successful. The character Strawberry Shortcake was created by Those Characters From Cleveland (TCFC), a collaboration between the licensing and character design branch of American Greetings cards and General Mills' Toy Group Marketing and Design Division. Strawberry Shortcake had first drawn attention as an illustration on an American Greetings card, but TCFC lifted her design from there and created a rag doll based on her likeness. The doll was popular, but not as great a success as they had hoped. Shortly afterward, they produced a kids' show and released more licensed merchandise based on Strawberry Shortcake and all the other characters on the show.[38] The show was so successful that profits made after its release quickly surpassed any profits they had made from the doll beforehand—the show became a one billion USD enterprise.[39] The success of Strawberry Shortcake demonstrated that advertisers had much to gain from selling products based on television properties rather than products that were only original concepts.

The story of *The Smurfs* (1980-1989) exemplifies the business behind children's television when character licensing first entered the scene. The limited liability company *La Société d'Edition, de Presse et de Publicité* (SEPP International S.A.) bought the rights to *The Smurfs* intellectual property from the original Belgian creator in the 1970s. It first created toys, which arrived on the market in the late 1970s, before producing a *Smurfs* television program that debuted on American networks in 1980 with great success.[40] *The Smurfs* show represented the synthetic nature of the new business model for children's television, wherein various aspects of *The Smurfs* franchise were handled by multiple independent companies working together. SEPP International S.A. owned *The Smurfs* intellectual property, but relied upon Hanna-Barbera to animate the show, while NBC broadcast the show on its affiliate network, and the Wallace Berrie Company created licensed products.[41]

While *The Smurfs* property represented the successful transition of an existing property to the new business model of licensed character television, *He-Man and the Masters of the Universe,* first broadcast in 1983, represented the shift towards greater concentration in the children's media industry as media and toy companies became more intertwined. *He-Man* was a joint project between the toy company Mattel and Filmation, an animation studio, co-financed by Group W (Filmation's parent company), and it was the first weekly series to be based purely on licensed characters that originated from the show itself.[42] *He-Man* is credited with being the "single major factor . . . for the immense new-production, kid-syndicated activity" because of how financially successful the close ties between companies were.[43] The companies behind He-Man were not merged yet—Filmation and Mattel were separate when making business decisions—but they were contractually obligated to each other, and their success encouraged more companies to follow suit and improve on the new formula.[44]

Finally, a great example of the end product of the push toward character licensing and risk management is *Thundercats.* This show was fully "toy/ television," where the companies involved—Telepictures (distribution/ syndication), Rankin-Bass (animation), Leisure Concepts Inc. (licensing), and LJN Toys Ltd.—actually shared profits from the show and the products, and all parties had a say in how the show was made.[45] This partnership was unprecedented in television: nowhere else on the television schedule did advertisers have as much influence/control over programming. Under the FCC regulations of the 1970s, as well as the newly accepted principles of reaching children with educational programming, such a development likely would have been banned by the FCC. During the Reagan era, however, television companies were able to make movements in the industry completely freely. In the case of both *He-Man* and *Thundercats,* almost everything that appeared on screen could be purchased in the nearest toy section, which was only possible because of the new business conglomerates.[46] This process of advertisers and television producers slowly merging over the time period of the 1980s shows how advertisers' primary goal of managing risk wildly changed the children's television industry.

For toy companies looking to manage risk, the next step was to produce their own shows to advertise products, and this resulted in almost the entire industry following in their footsteps in pursuit of drastically increased profits. At the beginning of the 1980s, non-product-based programs were quickly outpaced by product-based shows, which then came into the majority on children's television. Shows like *Fat Albert* and *Captain Kangaroo* ended at the beginning of the 1980s, likely prematurely.[47] But at the same time there was exponential growth in product-based programs, from zero shows in 1980, to fourteen in 1982, and then forty in 1984 across all networks originating from many different companies.[48] Producers discovered that television could lengthen the shelf life of a toy for years and could create new interest in old properties with reruns or renewals.[49] One example of this was the realization that, when a new season of a show was released, consumers bought toys created for the first season along with the new toys based on the second season. As a result, there was extreme growth in the number of licensed product sales over the 1980s.[50] Between 1983 and 1988 alone, revenue from product-based shows rose from 26.7 billion USD to 64.6 billion USD.[51]

The rapid expansion of media for children along with the focus on profits led to weaker, lower-quality shows. Many of these shows have been described as "cookie-cutter."[52] Cookie-cutter shows offered rough animation and repetitive plots. In terms of animation, many shows reused extremely repetitive backgrounds and animations, and incorporated low-quality special effects. Plots were similarly repetitive and reused. Consider the general plot of an episode of *He-Man:* a group of good guys, led by a masculine hero (muscular, solving problems through physical power, clever and witty, confident, unfazed by danger, with royal heritage) who uses a magic sword, must stop the group of bad guys that are up to no good. Now consider the general plot of *Thundercats*: a group of good guys, led by a masculine hero who uses a magic sword, must stop the group of bad guys that are up to no good. These trends were not limited to the most popular shows either, they extended much farther into less popular shows. In another blow to quality, some studios were even less artful in creating programs to advertise to children: for example, NBC's *Adventures of the Gummi Bears* was a children's television program that was created to sell gummy bears of the exact same brand name.[53] One critic described it as the "kidvid ghetto," where production companies and programmers only cared about "high-profit, quick turnover" and deemed it the "most exploited sector of the market."[54]

One 1994 survey noted that seventy-two percent of parents were frustrated that Saturday morning cartoons felt like one long commercial and ninety percent of those parents were concerned their children were becoming too materialistic.[55] Many saw the technology of television being used to sell to children as having moved so fast that the common person, not to mention government regulations, could not keep up.[56] There was little art in shows created to sell something so blatantly: shows that followed a cookie-cutter model possessed little artistic or intellectual merit. Deregulation of television opened the medium to exploitation as a hedge against corporate risk. The resulting programming was designed for little more than corporate profit, introducing characters hardwired for marketability and flooding every market that could feasibly apply to a child with those characters. The increased attention and care given to corporate risk management operated to limit the communicative potential of television: education and democratic access to information and general quality fell away in the rush for profits. This decrease in quality shows another way in which risk management changed the children's television industry.

DIVIDE AND CONQUER: THE MARGINALIZED THROUGH ADVERTISERS' EYES

At the time that *Sesame Street* was created, educators, policymakers, activists, and media professionals perceived in television the potential for greater inclusion through representation of the diversity of American society. In an environment mired in the fight over important civil rights issues, such as the challenge to redlining in 1968 and the ongoing fight for desegregation, television was pivoting towards attempting to show minority groups like Black/African Americans, American Indians and Alaskan Natives, Asian Americans, and Hispanic/Latin Americans in fairer lights.[57] In a 1951 study, for example, Black children demonstrated they were often uncomfortable with their own skin color: when asked to choose between a white and a black doll, Black children chose the white dolls a majority of the time.[58] A new generation of shows like *Sesame Street* (1969), *Fat Albert and the Cosby Kids* (1971), and *The New Mickey Mouse Club* (1976), expanded racial representation on television and helped kids gain a better understanding of themselves and other racial groups.[59] This progress was cut short in the 1980s when the goals and intentions of broadcasters aligned with advertisers more than those interested in social progress.

By the 1980s, representations of various racial groups were much more complicated. They were not necessarily represented positively, but they often were not represented solely negatively. When it came to live-action television, Black people were often depicted as hostile to members of their own group (especially couples) while acting very friendly towards European Americans. This reinforced the stereotype of infighting and black-on-black violence.[60] Animation was no different from live-action. As the dominant mode of children's entertainment, animation provided extra control for producers over the number and representation of people of color in a show. No matter who voiced a character, they could be animated however show runners wished, including in ways that completely avoided having to prescribe real-life races to each character.

The new advertiser-producer conglomerates did not include many people of color's voices, if at all. This is exemplified well by the struggles of Native Americans (including Alaskan Natives), who have very often been depicted as second-class citizens. Researchers have shown that both Native and non-Native children were aware of such prejudice because of television programming: when asked, children regardless of race spoke negatively about American Indians and Alaskan Natives.[61] Historically, Asian Americans were often represented as the "model minority." As such, they were often the target of "controlling images" in the media—limiting the representation of Asian American characters to minor supporting (and radicalized) roles such as the sidekick, a female newscaster, a benign mystic, etc.[62] All of these depictions were deceptively "normal" – reproducing social attitudes in ways that might not be noticed by the viewer, but kids who watched large quantities of this content would pick up on these signals of implicit inferiority.[63]

During commercial breaks on children's television, public service ads with the goal of often appeared meant to teach the kids various life lessons. Ads in the public service were quite different from commercial advertisements: prescriptive public service ads usually included a diverse cast of kids, including Black kids, and they were set in urban areas. By contrast, ads for products almost always only depicted white people and were usually set in rural locales. By doing this, programmers associated racial minorities with urbanness, problems, and even impurity, while on the surface appearing to be espousing good morals.[64] This radicalized vision of America resulted from implicit bigotry and reflected contemporary social fears, passing them down from one generation to the next through tropes of television and other types of storytelling. Moreover, such ads fulfilled the economic goal of reducing risk: by enforcing such stereotypes, advertisers were able to predict children's reactions more consistently to particular types of commercials, and more easily sell products to them.

One may think these conglomerates were foolish for ostracizing and alienating possible consumers, but this was a calculated move on the part of broadcasters, for which the primary product was their audience. Broadcasters sold audiences to advertisers, which used particular content to drive sales of products to audiences. Television had the ability to construct meaning by presenting a unique vision of reality to kids, who consistently would take that reality as truth.[65] Essentially, American television sold "normal life" to audiences and, for broadcasters, TV reflected not just anyone's life, but the lives of white middle-class Americans. Consider what those corporations knew: the effects of more than a century of racist policies that prevented swaths of people from accumulating generational wealth were completely unmitigated and the Reagan administration was completely uninterested in re-balancing the system.[66] The people who reliably had the most money were the white middle class. To the corporations, shaping, then pandering to, what most white people believed was the most profitable and risk-averse decision possible—any consequences to that decision were negligible.

This is another reason why the advertiser-producer conglomerates so entirely enveloped the industry (apart from public broadcasting): they needed to present a united front for their audience/products. This sort of ideology is reflected in the character licensing space, in which characters existed as product and nothing more. For example, when licensing groups bought the rights to *The Smurfs* or *Pac-Man,* the original creators no longer had any say in what the license companies made in pursuit of profit.[67] This limited the kinds of representations available. As an example, American Indians and Alaskan Native creatives were consistently blocked from creating original media, especially for children, mainly by being passed over in favor of other creatives. Television programming has been an inherently creative process, but producers who saw children's programming's purpose as purely economic sought particular kinds of programming with a particular point of view. If these independent creators, were allowed into the industry, then it might challenge the unified force these groups put forward, complicating the market space.

There are many examples of the ways businesses managed their risks by pandering to the perceived desires of the white middle-class population, which were shaped in part by the political culture of the Cold War. One of the most common of the Asian stereotypical roles of the postwar period was the Asian orphan, used for propaganda against the communists in Asia. This exploited white people's humanitarian instincts without acknowledging the role of the United States in the wars that created these orphans.[68] As another example, Latinos and Hispanics were widely ignored in American-produced television. The vast majority of content broadcast to those who spoke Spanish was broadcast or imported from Mexico.[69] Where American companies did invest in Spanish-language content, which was not that common, they simply dubbed over pre-existing shows (a significant outlier was *Sesame Street,* once again).[70] Not investing in Spanish was a conscious decision by companies, as they believed most Latin/Hispanic people would be bilingual by default.[71] Cultural attitudes reinforced by the principles of risk management dictated that the lost audience from people who could not speak English would not hurt their bottom line.

Boys vs. Girls: Cosmetic or concrete?

Beyond race and ethnicity, views on gender also heavily affected children's television in the name of risk management, namely through the distinct separation of boys' and girls' television. In the 1960s and 1970s, representation of girls was not good though progress was slowly being made. Superhero shows were among the first to portray empowered females, such as in *Birdman and the Galaxy* and *The Fantastic Four,* but it was not until 1970 that the first girl with supernatural powers was the main character in a series, in this case, *Sabrina the Teenage Witch.*[72] In 1974, *Sesame Street* aired an episode that featured a female construction worker; such a representation was so new that it came under fire from parents, but *Sesame Street* had a strong enough position to rebut those criticisms.[73] A 1974 study analyzed the top ten of the most popular children's programs and found that fewer than half incorporated a single female character.[74] However, even this little progress came to a halt in the deregulated 1980s. In 1982, another study found that there were three times as many men as women on children's television.[75] Clearly, with such minimal progress made, groups in the industry were likely maintaining a division.

This division persisted because advertisers believed that most of society believed girls and boys were fundamentally different and that it would be risky to challenge those ideas. Even in commercial ads there was a significant difference between ads aimed at different genders. Researchers found that ads intended for boys focused on practical/physical ideas; ads aimed at both boys and girls (though uncommon) focused on tradition and some emotional ideas; and ads intended for girls focused on highly emotional ideas.[76] If that sounds familiar, perhaps it is because it also describes shows like *GI Joe* and *Strawberry Shortcake*. When advertisers took over broadcasting for children's television, they began to apply the same ideas about what kinds of toys kids would buy to what they would watch. The conviction that girls would watch content designed for boys, but boys would not watch content made for girls, emboldened advertisers to create far fewer television programs aimed at girls.[77] Rather than invest in content that could fairly appeal to both genders and push the medium forward, advertisers chose to stick with what they believed managed their risk the best, which lined up with sexist ideas of the day.

The segregation of girls in children's programs and ads led girls to believe they were the undesirable, secondary gender and must embody similar characteristics to those depicted in representations of girls onscreen.[78] In *The Smurfs*, for example, there were ninety-nine male Smurfs embodying a very wide selection of personalities, along with Papa Smurf, but there was only one girl, Smurfette.[79] Smurfette is often treated as an object of desire by the other Smurfs, though she remains single and unattainable. Not only is she different from the other Smurfs by not having a descriptor attached to her name, but she is also alone in her gender. This all speaks to the idea that while a girl's presence is "enjoyed," she is more of an object and not really a part of the group on the same level.

In another example of secondariness, producers introduced She-Ra as a character in *He-Man* after they noticed lots of girls watching the show, though it originally had been conceived as a show for boys.[80] While the character was portrayed as "female," the show did not differentiate her from He-Man in any meaningful way, other than aesthetic differences. While she became popular, it still reinforced the secondary characteristics of women in these 1980s kid's shows. Both Smurfette and She-Ra exemplify the concept of secondariness that perpetuated widely held but unfounded expectations about "boys" and "girls."

While most programming was tailored to boys, more ads were directed at girls, often creating or contributing to unrealistic and dehumanizing conventions of "beauty." Television intentionally depicted impossibly beautiful people, which negatively affected girls' perceptions of themselves.[81] In order to meet those higher standards, girls would inevitably turn to products they saw on the same television programs. Unsurprisingly then, one study found that for every sixteen "boy" ads there were thirty-six "girl" ads; many of the extra ads sold cosmetic, beauty, and clothing products (some with character licensing attached).[82] This coordinated movement, depicting "pretty" girls and then selling ways to be "just like them," was easily orchestrated by the advertising conglomerates due to their chokehold over the industry. Contemporary researchers asserted that "they restrict opportunities for individual growth; they bias our ability to see males and females as individuals."[83] So in the end, alienation of the genders in children's media served multiple purposes—companies could aim products towards one group or another more easily while ensuring that girls would spend even more than boys.

CONCLUSION

The world of children's television is wildly different in the present day, and yet the conventions, practices, ideologies, and principles of this era are still apparent. Newer regulations have made program-length commercials a thing of the past, but product placement continues: in just the past few years audiences could attend a new Strawberry Shortcake movie (2021) and, in 2022, a new *Transformers* show appeared on Paramount television.[84] Historically, toys often had been inspired by books, films, and television, but it was not until 1984 that the licensed products of children's television completely dominated the market.[85] They do not dominate as much anymore, the regulations on such blatant advertising within a show have done their job, but the shows still persist. And what lessons have been learned from the way representation is used in children's television? *The Owl House* is a series lauded for its broad representation of queer characters, which Disney has been proud to advertise, but behind the scenes, the creator had to fight tooth and nail against Disney to bring that representation to the screen.[86] There's no good answer for how much progress has actually been made, but the 1980s offer lessons and insights for analyzing the landscape of children's television today.

This chapter aimed to analyze the era of children's television during the Cold War, focusing on the 1980s as an important period that reshaped what was available on television. Due to its widespread adoption, television viewing was marked as the most universal childhood experience of this time period.[87] With far-reaching implications because of the universality of children's television, the actions of advertisers, broadcasters, producers, and more "spoke" to particular audiences in cultural shorthand that perpetuated consumerist ideas and racial and gendered stereotypes and passed them on to the next generation. Children's television was a means to an end for the media/toy conglomerates of this time; one critic lamented that "technical prowess has passed our social and political abilities." [88] If parents are not discerning with the media they and their children consume, if society allows itself to be defined by what corporations believe it believes, this will easily happen again. The echoes of this time period are felt strongly throughout children's television, and it is possible it will never fully recover despite the great progress made in the past few decades. That could all be undone in the right conditions, so it is important to remember where you have come from in order to inform how you react to the present.

Notes

1. Robert W. Morrow, *Sesame Street and the Reform of Children's Television* (Baltimore: Johns Hopkins University Press, 2006), 83.
2. Jim Warren, "Children's Television Workshop," *Educational Researcher* 5, no. 8 (1976): 7.
3. Warren, "Children's Television Workshop," 7; J. Alison Bryant, *The Children's Television Community (*Mahwah, N.J.: Lawrence Erlbaum Publishers, 2007), 24.
4. Morrow, *Sesame Street,* 3.
5. Morrow, *Sesame Street,* 51.
6. Ignacio L Götz, "On Children and Television," *The Elementary School Journal,* Vol. 75, no. 7 (1975): 417. Morrow, *Sesame Street,* 66.
7. Kenneth Jost, "Children's Television: Will the new regulations make it better?" *CQ Researcher*, Vol. 7, no. 31 (1997): 728.
8. Warren, "Children's Television Workshop," 6.
9. Bryant, *The Children's Television Community,* 16.
10. Ashley Rodriguez, "Watch: The First TV Commercial, Which Aired 75 Years Ago Today," Quartz, G/O Media, July 1, 2016. https://qz.com/721431/watch-the-first-tv-commercial-which-aired-75-years-ago-today.
11. Nan Gray, and Sylvia Sunderlin (eds.), *Children and TV: Television's Impact on the Child (*Washington, D.C.: Association for Childhood Education International, 1967), 3.
12. Gray et al., *Children and TV,* 33.
13. Gray et al., *Children and TV,* 36.
14. Critic Les Brown cited in Morrow, *Sesame Street,* 2.
15. Bryant, *The Children's Television Community,* 12.
16. Bryant, *The Children's Television Community,* 12.
17. Bryant, *The Children's Television Community,* 13.
18. Bryant, *The Children's Television Community,* 7.
19. Allen K. Rostron, "Return to Hot Wheels: The FCC, Program-Length Commercials, and the Children's Television Act of 1990," *Hastings Communications & Entertainment Law Journal*, Vol. 19, no. 1 (1996): 60.
20. Bryant, *The Children's Television Community,* 16.
21. Bryant, *The Children's Television Community,* pg. 17-18.
22. Rostron, "Return to Hot Wheels," 61-62.
23. Rodriguez, "Watch: The First TV Commercial."
24. Information about the mute button and VCRs from Todd Gitlin, *Watching Television: A Pantheon Guide to Popular Culture*, 1st ed. (New York: Pantheon

Books, 1986), 107-108.

25. Bryant, *The Children's Television Community,* 17.
26. Bryant, *The Children's Television Community,* 17.
27. Norma Odom Pecora, *The Business of Children's Entertainment*, The Guilford Communication Series. (New York: Guilford Press, 1998), 7.
28. Cyndee Miller, "Parents tend to buy what their children see on TV," *Marketing News*, 28, no. 3 (1994): 2.
29. Gitlin, *Watching Television,* 104.
30. "Television Advertising Leads to Unhealthy Habits in Children; Says APA Task Force," Press Releases 2004, American Psychological Association, 2004, https://www.apa.org/news/press/releases/2004/02/children-ads.
31. Gitlin, *Watching Television,* 72.
32. Pecora, *The Business of Children's Entertainment,* 46.
33. Pecora, *The Business of Children's Entertainment,* 49.
34. Pecora, *The Business of Children's Entertainment,* 49-50.
35. Erica Wagner, "Peter Rabbit Blazed a Trail Still Well Trod," *The Times & The Sunday Times*, December 23, 2009.
36. Bryant, *The Children's Television Community,* 18.
37. Gitlin, *Watching Television,* 71.
38. Gitlin, *Watching Television,* 71-72.
39. Rostron, "Return to Hot Wheels," 61.
40. Pecora, *The Business of Children's Entertainment,* 65-66.
41. Pecora, *The Business of Children's Entertainment,* 67.
42. Pecora, *The Business of Children's Entertainment,* 69.
43. Pecora, *The Business of Children's Entertainment,* 68; 72.
44. Pecora, *The Business of Children's Entertainment,* 72-73.
45. Pecora, *The Business of Children's Entertainment,* 74-75.
46. Pecora, *The Business of Children's Entertainment,* 70; 75.
47. Bryant, *The Children's Television Community,* 7.
48. Pecora, *The Business of Children's Entertainment,* 61.
49. Pecora, *The Business of Children's Entertainment,* 51.
50. Pecora, *The Business of Children's Entertainment,* 57.
51. Rostron, "Return to Hot Wheels," 61.
52. Gitlin, *Watching Television,* 69.
53. Gitlin, *Watching Television,* 85.
54. George Gerbner, "Children's Television: A National Disgrace," *Pediatric Annals* 14, no. 12 (1985): 826.

55. Miller, "Parents tend to," 2.

56. Gray et al., *Children and TV*, 38.

57. CollegeHumor, "The Disturbing History of the Suburbs" *Adam Ruins Everything*, *YouTube,* 4 October 2017, Educational Video, 0:48–4:00, https://youtu.be/ETR9qrVS17g.

58. Henry Jenkins (ed.), *The Children's Culture Reader (*New York: New York University Press, 1998), 14. https://virginiatech.on.worldcat.org/oclc/39223599.

59. Bryant, *The Children's Television Community,* 14.

60. Gordon L. Berry and Joy Keiko Asamen, *Children & Television: Images in a Changing Sociocultural World* (Newbury Park, CA: Sage Publications, 1993), 182.

61. Berry and Asamen, *Children & Television*, 197.

62. Berry and Asamen, *Children & Television*, 205–206.

63. Gitlin, *Watching Television,* 80.

64. All information regarding public service ads from Gitlin, *Watching Television,* 99; 104–105.

65. Berry and Asamen, *Children & Television*, 186.

66. CollegeHumor. "The Disturbing History of the Suburbs," 0:48–4:00.

67. Gitlin, *Watching Television,* 74.

68. Berry and Asamen, *Children & Television*, 211.

69. Berry and Asamen, *Children & Television*, 222.

70. Berry and Asamen, *Children & Television*, 225–226.

71. Berry and Asamen, *Children & Television*, 224.

72. Bryant, *The Children's Television Community,* 13.

73. Warren, "Childrens' Television Workshop," 7.

74. Alice Sterling Honig, "Television and Young Children," *Young Children* 38, no. 4 (1983): 70.

75. Honig, "Television and Young Children," 71

76. D. W. Rajecki, Jill Ann Dame, Kelly Jo Creek, P. J. Barrickman, Catherine A. Reid, and Drew C. Appleby, "Gender Casting in Television Toy Advertisements: Distributions, Message Content Analysis, and Evaluations," *Journal of Consumer Psychology (Elsevier Science)* 2, no. 3 (January 1, 1993): 320.

77. Janet Wasko and Eileen R. Meehan, (eds), *A Companion to Television* (Newark: John Wiley & Sons, Incorporated, 2020), 335.

78. Frederick Williams, Robert La Rose, and Frederica Frost, *Children, Television, and Sex-Role Stereotyping* (New York, N.Y.: Praeger, 1981), 8.

79. Pecora, *The Business of Children's Entertainment,* 65.

80. Pecora, *The Business of Children's Entertainment,* 69.

81. Berry and Asamen, *Children & Television*, 198.

82. Dame Rajecki et. al., "Gender Casting in Television," 314.

83. Williams, Rose, Frost, *Children, Television, and Sex-Role Stereotyping,* 2.

84. "Strawberry Shortcake: Berry in the Big City," IMDb.com, September 18, 2021, accessed April 23, 2023. https://www.imdb.com/title/tt7996132/. "Transformers: Earthspark," IMDb.com, November 11, 2022, accessed April 23, 2023. https://www.imdb.com/title/tt14298658/.

85. Bryant, *The Children's Television Community,* 165.

86. Alex Dudok De Wit, "Disney Executive Tried to Block Queer Characters in 'The Owl House,' Says Creator," *Cartoon Brew*, August 14, 2020, accessed April 23, 2023. https://www.cartoonbrew.com/disney/disney-executives-tried-to-block-queer-characters-in-the-owl-house-says-creator-195413.html.

87. Wasko and Meehan, *A Companion to Television,* 327.

88. Gray et al., *Children and TV*, 38.

Civil Rights, NFL Television Culture and Integration

DANIEL SMITH

When World War II ended, countless Americans who had served in the Armed Forces returned home to the United States. The economy was booming, and professional sports like the National Football League began to surge in popularity.[1] Many service members were greeted with open arms, as they had just fought against tyranny and oppression in Europe. But the postwar euphoria did not extend to all equally and some soldiers were hardly recognized at all. More than one million African Americans had served in WWII, almost entirely in segregated units, only to return to a nation that broke its promise of equal rights for all of its citizens. Despite fighting for their country, America was not fighting for them. One such African American, George Dorsey, returned home after serving for five years in the U.S. Army, in 1946. He was not greeted with open arms. Instead, he and three of his companions, all Black, were brutally murdered in a coordinated attack.[2] George Dorsey's only crime was having been born African American. This is but one example of the racial violence experienced by African Americans in the postwar years, in which African Americans were often targeted by a white mob, and often killed.[3] The United States had fought fascism in Europe to promote the spread of freedom and democracy abroad, yet home-grown racism still gave white mobs license to target and kill African Americans, often with little more than a slap on the wrist from the justice system.

As American racism persisted in a postwar world more attuned to the issues of liberty, equality, and justice, the U.S. found itself in an ideological war with the Soviet Union. The Soviets capitalized on the postwar movement for decolonization to "liberate" nations under western colonial rule and encourage the spread of communism across Asia and Eastern Europe. The Soviets offered a vision of liberty, equality, and justice built on communist ideals, and used Eastern Europe as a sphere of influence for communism to develop. The United States, the premier example and main driver of the movement for liberal capitalist democracy, was meant to be "distinguished by free institutions, representative government, free elections," and most notably, "freedom from political oppression."[4] Yet, criticisms from abroad, including from the Soviet Union, on the unequal treatment of African Americans resulted in "anti-American" propaganda. Responding to such criticism, President Harry Truman realized that the promotion of human rights would begin at home and called for correcting "the remaining imperfections in our practice of democracy."[5] If the United States were to improve its appearance and appeal on the world stage, drastic changes had to be made to its domestic policies. Accordingly, the Truman administration began to "adopt a pro-civil rights posture" to promote democratic values. Moreover, the administration took this a step further by promulgating the Truman Doctrine and providing "political, military and economic assistance" to democratic nations under threat from communist influence.[6]

Events at home continued to thwart the U.S. government's struggle to improve its global record on civil rights in the face of the Cold War. In grappling with how to address racial tensions in postwar America, the federal government faced resistance from a number of states that were still committed to the Jim Crow regime. Particularly in Southern states, leaders of institutions such as universities or military bases were more opposed to equality for African Americans and actively worked against integration. The

Truman administration enjoyed limited success on civil rights by addressing segregation in the U.S. Armed Forces, but ending segregation in other facets of American life, such as education, would not occur for several more years.

In the aftermath of the war, Americans were spending more time at leisure, attending sporting events, for example, and professional football games in particular. The National Football League flourished the most in postwar America, as attendance numbers soared and household television viewership exponentially increased, further contributing to its explosion in popularity. The NFL also began to integrate its teams, after more than a decade of prohibiting the signing of African Americans to their rosters.

Players in the modern-day NFL are predominantly African American, and many Black stars like Patrick Mahomes are household names. Super Bowl LVII (57), played in 2023, marked a new milestone in NFL history: two African American starting quarterbacks faced off for the first time in the history of the Super Bowl. In the years before and during World War II, however, that ideal world where the Black athlete could play in any professional sport without regard to his or her race was almost impossible to imagine. A select few African Americans had played in the NFL's early days; for example, Fritz Pollard was the first Black person to play and coach in the NFL in the 1920s.[7] In the 1930s, however, the NFL silently stopped signing "negroes" to their rosters, ending the hopes of many collegiate star, who had hoped to continue playing professionally. Washington Redskins owner, George Preston Marshall led the effort to exclude African Americans from the sport.[8] The NFL remained a "whites only" sport until 1946.

This chapter argues that the NFL played a crucial role in the story of integration as well as the broader Civil Rights Movement, bolstered by its participation in the television culture of postwar America. I will briefly describe the steps taken by the federal government in response to criticisms abroad regarding its discriminatory policies against African Americans in the early postwar era. I also describe the emergence of the NFL as one of the most popular pastimes in postwar America, which it did in part by embracing the power of television and rejecting the segregation regime that plagued the southern states. By the time the Truman administration began to address segregation in the military, the NFL's integration process was well underway. Between 1946 and 1961, the NFL was thriving with the explosion of the television industry during the 1950s, and every team except the Washington Redskins had integrated. When the Kennedy administration took over in 1961, the Redskins found themselves front and center in a showdown in the nation's capital on the issue of integration in the broader scope of civil rights.

RACE: AMERICA VS. THE WORLD

In the postwar era, the first sign of meaningful progress on civil rights came on July 26, 1948, when President Truman signed an executive order that banned segregation in the United States Armed Forces. Executive Order 9981 stated that any American male was permitted to serve in the U.S. military "without regard to race, color, religion or national origin."[9] While it was an important statement of principle, the United States military still fought the first armed conflict of the Cold War—the Korean War (1950-1953)—with mostly segregated units, at least initially. Desegregation only came to pass when it became apparent that the dwindling number of American forces were making the war effort difficult to maintain. In 1950, during the Battle of Pusan, the United States military called up "reinforcements without regard to race," which was an important milestone in the fight for equality for African Americans.[10] By the time the Korean War Armistice was signed in 1953, over 600,000 African Americans were serving in integrated units.[11]

Despite the success of integrating the U.S. military during the Korean War, segregationists persisted in their protests against desegregation in other areas of American life, such as public schools, restaurants, and even restrooms. They even used the language of liberty to defend their positions, connecting their interests to the resurgent anti-communist movement in the U.S. At the time, many segregationists defended segregation as a fundamental part of white American values, claiming that to abolish segregation would align more with communism and "undermine the fabric of American society."[12]

Indeed, anti-communism in the U.S. quelled civil rights activism in the immediate postwar period, delaying further social progress.

But civil rights leaders made progress through the courts. The Supreme Court ruling on *Brown v. Board of Education* in 1954 unanimously ruled that segregation of public schools violated the "Equal Protection Clause" of the 14th Amendment and was therefore unconstitutional, mandating the desegregation of American public schools.[13] The ruling faced a movement of "massive resistance" in the South, delaying the school desegregation process for several years; moreover, the decision did little to desegregate other aspects of American life for African Americans, whether it was attending a college of their choice, going out to restaurants or theaters, or playing professional sports. It still took another decade before American society began to deliver on the promise of equal rights for African Americans.

When President John F. Kennedy took office, civil rights became a priority for his administration. The deep South remained heavily opposed to the desegregation of its schools and society as a whole, but President Kennedy used an incident at the University of Alabama in 1963 to address this issue head on. In his Address on Civil Rights President Kennedy acknowledged that the National Guard of Alabama were involved in maintaining a peaceful atmosphere in order for two Alabama residents that were accepted into the university to be admitted, despite having been born African American. No issues occurred, and the two students were peacefully admitted on campus, with the other students at the university behaving responsibly. President Kennedy called for every American to take a step back and think about their opinions on race. He said "And when Americans are sent to Vietnam or West Berlin, we do not ask for whites only. It ought to be possible, therefore, for American students of any color to attend a public institution they select without having to be backed up by troops."[14] Though Kennedy did not live to see it, the Civil Rights Act of 1964 and the Voting Rights Act of 1965 dismantled Jim Crow. It broke down barriers in the workplace and enabled African Americans and women greater access to opportunities such as higher education, especially at Southern universities.[15]

AMERICAN FOOTBALL AND TELEVISION

While the United States government continued grappling with its global image on civil rights, Americans at home were more focused on their own personal interests, and they turned to sports, among other leisure pursuits, to escape the grim reality of a Cold War and a murky record on social justice. Professional sports, primarily Major League Baseball (MLB) and the National Football League (NFL), had been a popular amusement in the United States since the early 1940s. This popularity, especially for the MLB, arose because it was the only mainstream sport played in America during WWII, and it helped to boost the morale of Americans on the home front.[16] Following WWII, the NFL surged in popularity and sought to generate even more growth.

The NFL had been fairly popular before WWII, but it was largely overshadowed by college football until the 1960s. In 1949, college football attendance for all teams, including teams not part of the National Collegiate Athletics Association (NCAA), was around 19.65 million. Attendance fell slightly in 1950 and continued to decline for the next few years, but still outpaced the NFL. In 1950, the NFL merged with the All-America Football Conference (AAFC) to create a thirteen-team conference that included the Cleveland Browns, San Francisco 49ers, and the original Baltimore Colts. Total attendance at NFL games that year was just 2,117,747.[17] Rising attendance across the 1950s was primarily due to conference expansion; the addition of the Dallas Cowboys in 1960 led to a jump in attendance to an average of 40,000 tickets sold per game. At the time, the NFL revenue stream relied heavily on attendance, as represented by ticket sales and gate receipts.[18]

Television was the key to the massive explosion in the postwar popularity of professional football. Televised broadcasts of the NFL actually date back as far as 1939, when television was not yet accessible or affordable to most

Americans, and sports coverage on television was in its infancy. The first televisions cost hundreds of dollars and were not yet mainstream, only sold to the wealthy few. Broadcasts were still very localized, and early sports broadcasts could not reach wide audiences with the limited technology available. In addition, the NFL owners were skeptical of telecasting games, which they believed had the potential to hurt attendance numbers, their main source of profit.[19] The New York Giants were the first team to experiment with televising games. They took the lead, signing a contract "with the National Broadcasting Company (NBC) to broadcast their home games for the 1947 season." New York City was "initially the epicenter of television" with nearly eight million residents. Because most television stations were found in cities, this gave NYC an advantage in televised games reaching its residents. The rest of the league observed from a distance. Owners remained divided on whether to televise their games, due to some noticeable drops in attendance during the 1948 season. For example, the Philadelphia Eagles had sold out only twice the previous season, although five different televised college games at Municipal Stadium had entirely sold out 70,000 seats. Concerned, the Eagles decided to ban telecasts of home games to encourage attendance. During the following season, the Eagles' attendance again dropped slightly, despite the team's stellar performance.[20] The overall consensus from NFL owners was that telecasting games did adversely affect gate receipts to some degree, but it remained difficult to provide "definitive evidence" on whether television hurt or helped attendance.[21]

In 1951, the league owners met to determine whether to allow broadcasts of all their games, home and away. The final vote was eleven to one in favor of allowing "telecasts in home territory with the permission of both the visiting and home clubs." One of the conditions for this was that visiting teams could broadcast their games anywhere in the country "as long as it did not interfere with any other team's home game or any other team's road game that was being televised or broadcast back to the home audience."[22] This policy was intended to encourage higher turnout at games. Some owners fully embraced this new medium, including the reviled racist at the center of the Washington Redskins integration showdown, George Preston Marshall, who "developed networks of stations" to broadcast home games outside of the seventy-five-mile radius from Washington D.C. that was "blacked out," (the games were not broadcast there).[23]

Within five years, it became apparent that television was the medium that would propel the NFL to new heights, reaching new and increasingly diverse audiences nationwide. In its 1956 contract with the NFL, the Columbia Broadcasting System (CBS) agreed to broadcast games across the country. CBS made use of its nearly 200 affiliated stations to telecast sixty-three NFL games that year, for eleven of the twelve teams. The first game broadcast on CBS was on September 30, 1956, between the Washington Redskins and the Pittsburgh Steelers. Bert Bell, the NFL commissioner at the time, said that it was "the greatest sports coverage in history," as it broadcast from East to West, from Maine to Seattle.[24] This led to an increase in publicity for the league, and the NFL was poised to be even more successful in the years to come.

A critical moment in football and television history was the NBC broadcast of the 1958 NFL Championship Game between the New York Giants and the Baltimore Colts. The game, which was the first and only NFL Championship to go to overtime before Super Bowl LI (51) in 2017, saw a whopping forty-five million television viewers tune in to watch the Colts defeat the Giants. The 1958 Championship Game has since been dubbed the "Greatest Game Ever Played" and helped spark the creation of a second league, the American Football League (AFL) one year later.[25] The 1958 Championship Game also marked a milestone in television history. In the immediate postwar years, televisions were incredibly difficult to obtain. They were labor-intensive to make and incredibly expensive for the time. For a ten–twelve-inch television in the late 1940s, one could expect to spend as much as $350.[26] Even so, televisions gradually became more affordable and popular throughout the 1950s. In 1946, only about 12,000 televisions were owned by Americans. This number exploded in the years following, and, by 1954, around 24 million televisions were in American homes; ownership continued to rise every year thereafter. Television had the challenge of finding something to broadcast on a given Sunday, and the solution was professional football. As the sport became more popular, television stations turned to the NFL as their Sunday afternoon entertainment, and the NFL reaped the benefits of millions of people tuning in to their game broadcasts.[27]

The NFL's success with television was not without criticism, especially when it came to blackouts. Leading up to the 1957 Championship game between the Detroit Lions and Cleveland Browns, critics challenged the blackout in Detroit, again meant to encourage ticket sales over watching the game from home. Despite backlash from the governor and a senator, both from Michigan, Commissioner Bell defended the decision: "I don't think it is honest to sell tickets to thousands of people, then afterward, when all the tickets are gone, to give the game to television."[28] He wanted people in the stands, not watching from home. The blackout policy was challenged once again five years later when the Giants hosted the Green Bay Packers in 1962, where again, the game was blacked out in New York City for the home team. This blackout policy was such an issue for the NFL that the first seven Super Bowls were not televised in the cities they were played in, even though the teams that played in the host cities were not on their home field.[29] It was not until 1973 that the controversial blackout policy was repealed. That year, Congress passed legislation that is still in effect today, which requires "any NFL game that was declared a sellout seventy-two hours prior to kickoff be made available for local TV."[30] The NFL continued to dominate the world of American sports through the 1960s despite such controversial blackout policies, and even as the owner of the Washington Redskins butted heads with the federal government over its lily-white policy.

NFL expansion was not limited to new markets and audiences. It also expanded access for African American players, who had been excluded from signing with NFL teams for two decades. This exclusion resulted from a "gentleman's agreement" among NFL owners in the early 1930s, instigated by Redskins owner George Preston Marshall as noted above. In spite of this, by the mid-1940s owners increasingly sought young talent from the collegiate level, including Black players who had made names for themselves at their schools and among the fanbase. Paul Brown, owner of the Cleveland Browns in the AAFC (Cleveland joined the NFL in 1950), signed Bill Willis and Marion Motley to the roster in 1946, even before the federal campaign for increased racial integration. Both players went on to Hall of Fame careers in the NFL. Brown's personal views on race are unclear, but he knew it was important to hire players based on talent, not race, in order for his team to be successful on the field. A brilliant football mind, Brown went on to have a stellar career in Cleveland, with a win-loss-tie record of 167-53-8, and he coached the team to three national championships in 1950, 1954 and 1955.[31] Arguably, this resulted from his decision to hire talented Black players. Brown served as the first and most significant example of an NFL owner who willingly integrated his team in the best interests of the league. He led the way in the integration of professional football after 1946. Most NFL teams followed suit in signing African American players.

Integrating NFL teams proved to be an issue complicated by personal beliefs, logistical issues, and the profit motive. Some owners understood that it was in the best interests of the league, while others needed convincing. Municipal and state approaches to desegregation helped shape their decisions. After the 1945 season, for example, a second Cleveland team—the Rams—moved to Los Angeles. There the lease agreement for the local football stadium, the LA Coliseum, required the team to sign African Americans to their roster. As a result of additional pressure put on the Rams by *Los Angeles Sentinel* sports reporter Halley Harding, the Rams reluctantly signed Kenny Washington and Woody Strode, each of whom had been star players at UCLA several years earlier.[32] But race laws varied around the country, which was still at the height of Jim Crow and segregation. In southern states Black athletes on the Rams and other NFL teams often could not stay in the same hotels or eat at certain restaurants with their teammates. In one such incident, the Cleveland Browns were supposed to travel to Miami to face the Dolphins, but Florida still enforced strict segregation laws, and two Black players for the Browns had to remain in Cleveland.[33] Even with the slow process of integrating the league following WWII, by 1960, twelve teams, including the brand spanking new Dallas Cowboys, had integrated their rosters. That is, every team except the Washington Redskins.

SHOWDOWN IN THE NATION'S CAPITAL

As the rest of the NFL integrated, the ownership of the Washington Redskins remained stubbornly opposed. Based in Washington, D.C., where there was an increasingly diverse population and a new administration under President John F. Kennedy that sought to address the civil rights issue on many fronts, the Redskins were about to become the center of national attention for all the wrong reasons. The team's owner, George Preston Marshall, had led the original effort to prohibit African Americans from playing in the NFL at the height of the Great Depression in the 1930s. Marshall embraced the nation's Jim Crow laws rather than join the rest of the NFL in breaking down barriers, and he felt no shame maintaining this position. Even as the rest of the league integrated, he continued to bar Black athletes from playing for his team. Marshall once claimed to reporters that the Redskins did not have Black players as a matter of circumstance: the team usually "recruited players from segregated Southern colleges." "Whites only" colleges meant the team did not have to worry about race in recruiting players. According to Marshall, the team roster's "lily-whiteness" was not due to his own prejudice. Instead, it was good business: the team's fans were predominantly from the South, where Jim Crow laws still heavily influenced what African Americans were allowed to do in American society. Indeed, until the expansion that introduced the Dallas Cowboys to the NFL, the

Redskins had been the team of the South, as no team was farther south except for Los Angeles and St. Louis. Marshall did not want to "offend" his Southern television and radio audiences by hiring and playing Black people.[34] At other times, however, his logic was more openly specious and defiant, suggesting his own racial prejudices. He once claimed that "We'll start signing Negroes, when the Harlem Globetrotters start signing whites."[35]

In 1961, a new administration under President John F. Kennedy took office, intending to pass legislation on civil rights issues. President Kennedy fervently believed in the desegregation of America, a belief that became the centerpiece of his domestic agenda. In his 1963 address to the nation on civil rights, Kennedy argued "It ought to be possible, in short, for every American to enjoy the privileges of being American without regard to his race or his color."[36] In particular, Kennedy was adamant about "ending racial discrimination in hiring," which held important implications for Marshall's policy of segregation for the Redskins. On March 6, 1961, President Kennedy signed Executive Order 10925, which established the President's Committee on Equal Employment Opportunity. In short, this committee required that contractors "take affirmative action to ensure that applicants are employed, and that employees are treated during employment, without regard to their race, creed, color, or national origin."[37] Interior Secretary Stewart Udall, hired by President Kennedy, took the lead in the effort to integrate the Redskins after finding violations in their hiring practices that conflicted with the lease for their new stadium.

Secretary Udall found the fact that the Washington Redskins were the last team in the NFL to not have signed a Black player to its roster to be "outrageous," and reasoned that it would be appropriate to "challenge the hiring practices of George Marshall."[38] Indeed, a week before Kennedy's executive order, Udall wrote to Marshall, calling out this behavior. Udall said he was "cognizant of the fact that there have been persistent allegations that Marshall's company practices discrimination in the hiring of its players." He followed this up by indicating he assumed that Marshall "will fully adhere to its contractual obligations."[39] Marshall's defiance was hurting his team and the NFL, as a whole, clinging to the dying Jim Crow era.

Secretary Udall provided briefings on the subject in his "Weekly Reports to the President" and offered a path to force Marshall to integrate his team. His office, the Department of the Interior, was building a new stadium for the Redskins. Udall proposed writing integration language into the lease: that the team must sign at least one Black player to its roster before they could use the stadium.[40] By mid-March, this news had caught the attention of local and state newspapers, and it quickly became a nationwide news story. Headlines such as "Redskins Must Use Negroes, Says JFK," or "Test in Capital: A Negro Plays or No Stadium," were written to capture the attention of the reader, which worked well for an issue such as this. With the eyes of America turning to Washington with the story, this undoubtedly put even more pressure on both Marshall of the Redskins and the Kennedy administration to come to a deal on the issue.

In the last week of March 1961, correspondence between Secretary Udall and Marshall leaked in local newspapers and were covered more broadly by national papers. Newspapers reported that Udall had accused Marshall of practicing "discrimination in the hiring of its players." In the correspondence, Udall described amendments to regulation policies under National Capital Parks, noting the amended regulations "prohibited discrimination in employment practices with respect to any activity provided for by a contract, lease or permit." The letter notified Marshall that he would have to comply with federal anti-discrimination policies in hiring for the Redskins or be prohibited from playing in the new stadium that was under jurisdiction of the National Capital Parks.[41] Marshall denied any wrongdoing in his letter responding to Udall. He declared that "the National Football League has no restrictions [on hiring that] I know of, [and] neither do the Redskins." He also argued that his team had not violated any laws and was compliant with the lease he had signed in 1959.[42] Indeed, when Marshall and the Armory Board were negotiating terms for the lease, the board had written in anti-discrimination requirements in hiring practices regarding use of the stadium. Marshall had taken no issue with these terms regarding the hiring of workers for the stadium itself, but he was still strongly opposed to hiring Black players for his team. As a result, the Armory Board had removed "the nondiscrimination clause relating to Redskins players."[43] The showdown in the nation's capital was inching closer to reaching its climax as the pressure continued to build against the Washington Redskins owner.

The showdown intensified, pitting D.C. area residents against racial extremists. Bobby Richards, a former high school football coach in the D.C. area, told *The Washington Post*, "My uncles and grandfather, they were against the Redskins," and continued saying that they would root for any team, such as Cleveland, "that had black players."[44] One E. B. Henderson wrote to the *Afro American* newspaper, urging both Black fans and whites to "boycott all Redskins games until the team integrated its roster."[45] On the other side, extremists and Neo-Nazis protested the idea of integrating the Redskins. Secretary Udall regularly received angry, if not threatening letters from white supremacists who were opposed the federal campaign to desegregate the United States. One letter from a Tennessee man used the N-word to describe an African American and claimed America was heading towards dictatorship for forcing the hiring of Black players.[46] Others played on the fear of communism, as the US was still in the midst of a Cold War with the Soviet Union. Segregationists used anti-communist sentiment to justify their racist tendencies in refusing to support the integration of the Washington Redskins.[47] All of this fueled the defiance of George P. Marshall, who only began to change his tone after the NFL Commissioner became involved.

In the face of Marshall's defiant position, Secretary Udall convinced NFL Commissioner Pete Rozelle to use his platform to put an end to this scandal that was plaguing the NFL and the nation's capital. Rozelle and Marshall met in August of 1961, whereafter Marshall seemed to be out of options for fighting the government or the NFL commissioner on the issue of integrating his team and conceded. They agreed that Marshall should plan on integrating his team "in the best interests of the sport." Marshall announced that his team would look to draft an African American player in the upcoming NFL Draft in December 1961.[48] On August 15 the *Albuquerque Journal* reported that Marshall had sent a letter to Rozelle confirming his intentions to hire a Black player. Secretary Udall decided to take Marshall at his word: he would allow the Redskins to use their new stadium of 50,000 seats with an all-white team under "the assumption that this offer has been made in good faith"—that is, that the team would sign Black players after the 1961 season.[49] But Marshall's reputation for racial prejudice preceded him: although the Redskins did draft Ernie Davis, Davis refused to ever "play for that S.O.B." and demanded to be traded. The Redskins traded Davis for the Cleveland Browns' Bobby Mitchell. Mitchell went on to a Hall of Fame career with the Redskins, playing with the team for seven seasons.[50] This marked the end of the showdown in Washington D.C., and the first time the National Football League had been fully integrated.

CONCLUSION

In the aftermath of this crisis, the National Football League continued to increase its audience by hundreds of thousands, if not millions, of viewers each year. Its sister league, the American Football League (AFL), also surged in popularity. The AFL television contract with NBC in 1965 led to a pitched competition between the leagues for "fans, players, and coaches," ultimately leading to talks about merging the two leagues. The announcement of a merger came in 1966, but the AFL officially merged with the NFL creating one league of operations in 1970, when they "integrated their regular season schedules."[51] Two conferences came out of this expanded NFL, which survive today: the American and National Football Conferences (AFC and NFC). The first Super Bowl between these conferences, originally referred to as the "AFL-NFL World Championship Game," took place on January 15, 1967, between the NFC's Green Bay Packers and the AFC's Kansas City Chiefs. The Packers easily defeated the Chiefs 35-10. Over the next several decades, the Super Bowl became a spectacle that continues to draw in tens of millions of viewers as one of the most watched sporting events in the world.[52]

By 1970, the three major television networks ABC, NBC, and CBS, were all poised to televise games for the newly expanded NFL, but it was ABC that pitched the idea that set the stage for NFL success for decades to come: *Monday Night Football.* ABC producer Roon Alredge (also featured in Jason Toy's contribution to this volume) was the mastermind behind *Monday Night Football* (MNF), and the NFL has thrived as a result. *MNF* first aired on September 19, 1970, a match-up between the New York Jets and the Cleveland Browns.[53] *MNF* remains vastly popular today, drawing in an average of fourteen million viewers each week in 2022. This popularity is despite the king of television, NBC's *Sunday Night Football*, having become the true prime-time event of the NFL since 2006, averaging a whopping twenty million viewers, also during the 2022 season.[54]

The NFL has remained a giant in the world of sports since the 1960s, and it likely will for years to come. Professional football is vastly different from what it was a century ago, after enduring scandals over hiring practices with regard to race and surpassing professional baseball as the American pastime. The success of television in the 1950s and beyond was arguably the biggest innovation that made the NFL the king of American sports. Aided with the well documented coverage of the NFL's integration process, the NFL became a sport that not just white men, but people of all races and sexes have come to enjoy each year.

Notes

1. David G. Surdam, *Run to Glory and Profits: The Economic Rise of the NFL during the 1950s* (Lincoln, NE: University of Nebraska Press, 2013), 58.
2. Mary L. Dudziak, *Cold War Civil Rights: Race and the Image of American Democracy (*Princeton, NJ: Princeton University Press, 2011), 18-19.
3. Dudziak, *Cold War Civil Rights,* 19.
4. Dudziak, *Cold War Civil Rights,* 27.
5. Dudziak, *Cold War Civil Rights,* 82.
6. Dudziak, *Cold War Civil Rights*, 27; "The Truman Doctrine, 1947." U.S. Department of State. U.S. Department of State. Accessed April 13, 2023. https://history.state.gov/milestones/1945-1952/truman-doctrine.
7. Steve Rose, "Racial Harmony in a Marxist Utopia: How the Soviet Union Capitalized on US Discrimination," *The Guardian,* January 24, 2016. https://www.theguardian.com/artanddesign/shortcuts/2016/jan/24/racial-harmony-in-a-marxist-utopia-how-the-soviet-union-capitalised-on-us-discrimination-in-pictures.
8. Surdam, *Run to Glory*, 281.
9. "Executive Order 9981: Desegregation of the Armed Forces (1948)" National Archives and Records Administration, accessed March 27, 2023. https://www.archives.gov/milestone-documents/executive-order-9981.
10. Dudziak, *Cold War Civil Rights,* 88.
11. "The U.S. Military Led the Way to Ending Segregation in America as African Americans Fought in Integrated Units during Korean War, " *The Seattle Medium*, September 24, 2013. https://seattlemedium.com/the-u-s-military-led-the-way-to-ending-segregation-in-america-as-african-americans-fought-in-integrated-units-during-korean-war/.
12. Dudziak, *Cold War Civil Rights*, 28.
13. "Brown v. Board of Education (1954)," National Archives and Records Administration. accessed March 27, 2023. https://www.archives.gov/milestone-documents/brown-v-board-of-education, "History - Brown v. Board of Education Re-Enactment," United States Courts, accessed March 27, 2023. https://www.uscourts.gov/educational-resources/educational-activities/history-brown-v-board-education-re-enactment.
14. John F. Kennedy, "Televised Address to the Nation on Civil Rights," televised Address to the Nation on Civil Rights, Columbia Broadcasting System, June 11, 1963. https://www.jfklibrary.org/learn/about-jfk/historic-speeches/televised-address-to-the-nation-on-civil-rights.
15. Warren K. Leffler, "The Civil Rights Act of 1964: A Long Struggle for Freedom Epilogue," Library of Congress, October 10, 2014. https://www.loc.gov/exhibits/civil-rights-act/epilogue.html.

16. Joey Corso, "Major League Baseball's Popularity during WWII," *Bleacher Report,* October 3, 2017. https://bleacherreport.com/articles/161265-major-league-baseballs-popularity-during-wwii.

17. "A Look at the NFL in the Fabulous 1950s." *USA Today,* October 11, 2019. https://www.usatoday.com/story/sports/nfl/2019/10/11/a-look-at-the-nfl-in-the-fabulous-1950s/40301077/; "Standings," Official Site of the National Football League, NFL.com. Accessed April 17, 2023. https://www.nfl.com/standings/league/1950/reg; "1950 NFL Attendance Data." Profootballreference.com, accessed April 17, 2023. https://www.pro-football-reference.com/years/1950/attendance.htm; "NCAA Division I Football Records (Attendance Records)." ncaa.org. Accessed April 17, 2023. http://fs.ncaa.org/Docs/stats/football_records/DI/2010/Attendance.pdf.

18. Surdam, *Run to Glory and Profits the Economic Rise of the NFL during the 1950s,* 58; Bob Carroll, "The American Football League Attendance, 1960-69." profootballresearchers.org, 1991. http://www.profootballresearchers.org/archives/Website_Files/Coffin_Corner/13-04-430.pdf.

19. Surdam, *Run to Glory and Profits the Economic Rise of the NFL during the 1950s,* 228-229.

20. All quotes regarding the Eagles in this paragraph from Surdam, *Run to Glory*, 230-232.

21. Surdam, *Run to Glory*, 231

22. Surdam, *Run to Glory*, 233.

23. Surdam, *Run to Glory*, 236.

24. Surdam, *Run to Glory*, 239; "The History of NFL Today on CBS: Mr. Pop Culture," Mr. Pop Culture, Exploring the World of Pop Culture, September 22, 2020. https://mrpopculture.com/the-history-of-nfl-today-on-cbs/.

25. The AFL eventually merged with the NFL in 1970. "A Look at the NFL in the Fabulous 1950s," *USA Today*, October 11, 2019. https://www.usatoday.com/story/sports/nfl/2019/10/11/a-look-at-the-nfl-in-the-fabulous-1950s/40301077/; "Colts Win NFL Title in 'Greatest Game Ever Played,'" History.com. November 30, 2021. https://www.history.com/this-day-in-history/nfl-greatest-games-colts-giants-1958; "NFL and AFL Announce Merger," History.com, November 16, 2009. https://www.history.com/this-day-in-history/nfl-and-afl-announce-merger.

26. Surdam, *Run to Glory*, 227.

27. Reed Karaim, "NFL Controversies," CQ Press Library, September 4, 2015. https://library-cqpress-com.ezproxy.lib.vt.edu/cqresearcher/document.php?id=cqresrre2015090400.

28. Surdam, *Run to Glory*, 240-241.

29. Al Yellon, "A Brief History of NFL Blackouts," SB Nation Chicago, September 14, 2010. https://chicago.sbnation.com/2010/9/14/1686025/a-brief-history-of-nfl-blackouts.

30. James Williams, "Time for the NFL to End the TV Blackout Policy," *Bleacher*

Report, September 20, 2017. https://bleacherreport.com/articles/1070123-time-for-the-nfl-to-end-the-tv-blackout-policy.

31. "Paul Brown: Pro Football Hall of Fame Official Site," accessed March 27, 2023. https://www.profootballhof.com/players/paul-brown/; Surdam, 286.

32. William Rhoden, "In 1946, Pressure from Black Citizens Forced the Los Angeles Rams to Desegregate," Andscape, February 10, 2022. https://andscape.com/features/in-1946-pressure-from-black-newspaper-reporter-halley-harding-forced-the-los-angeles-rams-to-desegregate/; Surdam, *Run to Glory,* 286.

33. Surdam, *Run to Glory*, 286-287.

34. Thomas G. Smith, *Showdown: JFK and the Integration of the Washington Redskins* (Boston, MA: Beacon, 2012), 153.

35. William Gildea quoted in "Integrating the Redskins: George Preston Marshall vs. the U.S. Government," *The Washington Post,* June 5, 2002. The Harlem Globetrotters were founded in 1926. They were and still are an exhibition basketball team that is well known for smashing social and racial barriers since the late 1920s, a fact belied by the defiant comment of Marshall. "About," Harlem Globetrotters, accessed April 13, 2023. https://www.harlemglobetrotters.com/about/.

36. John F. Kennedy, "Televised Address to the Nation on Civil Rights," CBS, June 11, 1963. https://www.jfklibrary.org/learn/about-jfk/historic-speeches/televised-address-to-the-nation-on-civil-rights.

37. "Executive Order 10925-Establishing the President's Committee on Equal Employment Opportunity." Executive Order 10925-Establishing the President's Committee on Equal Employment Opportunity | The American Presidency Project, March 6, 1961. https://www.presidency.ucsb.edu/documents/executive-order-10925-establishing-the-presidents-committee-equal-employment-opportunity.

38. Smith, *Showdown*, 150.

39. Stewart Udall, "File #3953: 'Letter to George P. Marshall, 1961,'" University of Arizona Special Collections, accessed April 23, 2023. https://speccoll.library.arizona.edu/online-exhibits/files/show/3953.

40. Smith, *Showdown,* 150.

41. Lewis F. Atchison, "Court Action May Follow Udall Warning to Marshall," https://speccoll.library.arizona.edu/online-exhibits/items/show/1437.

42. George Preston Marshall, Letter to Stewart Udall, Interior Secretary, Washington, District of Columbia, March 24, 1961, https://speccoll.library.arizona.edu/online-exhibits/files/original/5897af3183a6e831df21c99b7142cc3a.jpg.

43. Smith, *Showdown,* 151.

44. Dan Steinberg and Chris L. Jenkins. "Black Fans Have Grown to Love the Redskins," *The Washington Post,* October 26, 2011. https://www.washingtonpost.com/sports/redskins/black-fans-have-grown-to-

love-the-redskins/2011/10/26/gIQA8q7EKM_story.html?tid=usw_passupdatepg.

45. Smith, *Showdown,* 154.
46. Smith, *Showdown*, 159.
47. Smith, *Showdown*, 160.
48. Smith, *Showdown*, 162.
49. "'Skins' Racial Conflict Apparently Cleared Up," *Albuquerque Journal,* accessed April 3, 2023. https://speccoll.library.arizona.edu/online-exhibits/files/original/becbdf49bb7adb8fded4af8c2fcbd8e1.jpg.
50. Shirley Povich, "George Preston Marshall: No Boredom or Blacks Allowed," *The Redskins Book*, hosted by *The Washington Post*, accessed April 26, 2023. https://www.washingtonpost.com/wp-srv/sports/redskins/longterm/book/pages/17.htm; "The Career of Bobby Mitchell," Pro Football Hall of Fame," https://www.profootballhof.com/news/2020/04/the-career-of-bobby-mitchell/, accessed April 26, 2023.
51. "NFL and AFL Announce Merger." History.com., November 16, 2009. https://www.history.com/this-day-in-history/nfl-and-afl-announce-merger.
52. *Ibid;* Dallas Robinson, "Super Bowl 1: Who Won the First Super Bowl in NFL History?" *Pro Football Network*, February 12, 2023. https://www.profootballnetwork.com/super-bowl-1-who-won-the-first-ever-super-bowl/.
53. Bill Hofheimer, "The Very First Monday Night Football Game Aired 42 Years Ago Today," *ESPN Front Row,* September 21, 2012. https://www.espnfrontrow.com/2012/09/the-very-first-monday-night-football-game-aired-42-years-ago-today/, "The Story of the First-Ever Monday Night Football Game," *Sports History Network*, August 26, 2021. https://sportshistorynetwork.com/football/nfl/first-ever-monday-night-football-game/.
54. Christina Gough, "NFL Monday Night Football TV Viewership 2022." Statista, May 23, 2022. https://www.statista.com/statistics/819112/nfl-monday-night-football-viewers/

The Best Trash of the Lot: Dallas, Melodrama, and 1980s America

CATHRYN JONES

The 1980s was a decade that upended the postwar liberal consensus and transformed American politics, society and culture. The "Great Society" inaugurated by the Johnson Administration in the 1960s was the centerpiece of "twenty years of political and social activism that expanded civil liberties" for a broad range of Americans.[1] The promise of greater equality and inclusion in the American dream did not survive the economic instability of the 1970s: the oil shock of 1973, persistent stagflation, and the ignominious end to the brutal war in Vietnam challenged Americans' faith in the Great Society.[2] In reaction to the civil rights and antiwar movements, moderate conservatives supported Republican leaders such as Nixon, who promised to restore 'law and order' in the face of liberal protests. A decade later, a new political regime under Ronald Reagan capitalized on this precedent, citing an apparent "alarming rise in the crime rate" to "intensi[fy] Nixon's 'war on crime" and justify a political program that turned away from the great society.[3] Reagan won the 1981 presidential election by promising to reinvigorate the American economy and put an end to social crises. In office, Reagan reframed the cause of criminality and passed deregulatory policies that drove "massive social inequality," while persistently invoking the desire to return the United State to a simpler, less controversial time.[4]

The less controversial time promised by Reagan was, as media and communications scholar Michael Dwyer describes it, the "fantasy fifties."[5] For Dwyer, this was "the seductive dream of 'the good old days,'" based on an imagined period of prosperity and unity of the first postwar decade.[6] "The Fifties" represented, for some people, a glorious time when America had achieved victory in World War II and was prosperous economically, before the humiliation of Vietnam and the social crises of 1970s economic decline. Calling back to that time of stability in the 1980s was a sort of escapism and an effective pathos appeal to voters in the 1981 election.

Television, too, was entering a transformative period in the 1980s. Broadcast television had saturated American homes, but new technologies of transmission and reception were reshaping the audience experience, unleashing what some have called a "golden age of television."[7] Cable television expanded the amount and type of available programming. For example, "Music Television" (MTV) began operating in 1981 and was a major cultural force that reshaped both the music and television industries in important ways. Competition pushed the networks to innovate their own schedules, which led to the rise of popular prime time soap operas like *Dallas*, and *Dynasty*, and new distribution networks made a number of them internationally known and popular as well.

Reagan-era values also found their way into popular culture and onto television screens. The role of the media in crafting and disseminating the Reagan era's nostalgia for the 1950s has been studied by many scholars. In their article "Reagan's Rainbow Rodeos: Queer Challenges to the Cowboy Dreams of Eighties America," for example, Christopher Le Coney and Zoe Trodd describe this same idea but focus on one aspect that they call "cowboy mythology."[8]

This cowboy mythology was a

> Reagan-era attempt to rewrite American defeat in Vietnam and silence criticisms of American involvement in that war. America entered the 1980s, and both Hollywood and the Reagan administration reclaimed [John] Wayne's legacy by reimagining the cowboy mythology that had been devastated during a "glory John Wayne war," as the Vietnam veteran Ron Kovic put it (158).[9]

This cowboy mythology, which glorified the rugged masculinity of cowboys and the nuclear family, connects back to the social crises that Reagan was invested in fixing. Promising to go back to "the Fifties" would fix the crises that had caused a general malaise in the 1980s. Calling back to "the Fifties" became a popular device in contemporary entertainment. In 1985, for example, Universal Pictures released *Back to the Future*, a movie in which the main character is accidentally transported to the world of his parents' teenage years; it was so popular it spawned two sequels. *Grease* glamorized the greaser of the 1950s with pop culture stars Olivia Newton John and John Travolta (*Welcome Back Cotter* and *Saturday Night Fever*). The pop star Madonna's "Material Girl" was a pop anthem for the materialistic 1980s; and in the video Madonna referenced Marilyn Monroe's musical number "Diamonds are a Girl's Best Friend" from *Gentlemen Prefer Blondes* through her dress choice and dance.

This chapter explores the role of *Dallas* in projecting and refracting the cultural values of Reagan-era nostalgia around rural and western settings.[10] I argue that *Dallas* became a cultural juggernaut in the 1980s by popularizing a new form of melodrama that blended the genre conventions of the western—uncritically nostalgic and traditional family values (which were also being espoused by Reagan)—with the conventions of daytime soap operas: melodrama and open narrative. As a "prime-time soap opera" the show reached large audiences made larger due to the nostalgic escapism of *Dallas*' western setting and emotionally realistic melodramatic plot.

By the 1980s, one of the more popular genres of entertainment television was the serial melodrama. Serial melodramas were defined by two traits: seriality and melodrama. Seriality denotes a narrative format: serials consist of successive episodes through which plots unfurl over time and narratives remain "open." Open narratives do not have a specific ending that the narrative is heading towards from the beginning and does not achieve "closure" at the end of an episode. This is opposed to closed narratives in forms like film or mini-series, which do have specific and realized narrative endings. Open narratives also tend to use cliffhangers at the end of their episodes, to entice their audience to return for the next installment.

Melodrama is notoriously difficult to define, but recent scholarship describes it as seeking two ends: "psychic realism and moral legibility."[11] Psychic realism is when the characters are relatable because, despite whatever unrealistic setting the characters may be in, their emotions and reactions feel correct to viewers. The characters feel real even when inside of "exaggerated conflicts between stock heroes or heroines and villains, evil intrigue, suspense, improbable plot twists, and happy endings," that have the goal of generating an emotional response from an audience.[12] Moral legibility is the other end goal of serial melodramas. Melodramas take on issues that are being discussed at the time of their creation, dramatize that moral problem, then offer a clear and understandable solution to it, or moral legibility.[13]

Scholars are still working to define "melodrama," but the public is clear on what it means: ridiculous and emotional. Popular colloquial uses of the term melodrama are usually insulting. Scholarly discussions of melodrama have also differed. Melodrama has not always been valued as an important topic of study; there was not a movement to take melodrama seriously until feminist scholars took it up in the 1970s.[14] However, once scholars began to study melodrama more seriously it came to be understood as a mode of communication that has a "symbiotic and complex relation to realism."[15] That complex relationship to realism is how melodrama achieves moral legibility; if the emotions are real, then the morals are understandable to viewers.

Serial melodramas were well known to audiences even before film and radio, and they were so popular that they were a common genre for new media, like radio and television at their beginnings, to gain financial viability.[16] A sub-genre of serial melodrama, soap opera, was especially popular among daytime television producers, who could fill airtime with cheap programming that served primarily as an "advertising vehicle for laundry detergents and household cleaning products," aimed at housewives.[17] The genre also appealed to their intended audience: women. Soap operas appealed to certain communities of women because they aired in the daytime slot, were comprised of repetitive and predictable story structure with an open narrative, and focused on women's stories. Each of these traits made soap operas appealing and easy to watch while doing domestic work.

Embedded in these genre conventions are certain assumptions about the structure of American society and the workday that were built into early television schedules and the programs that filled them. The daytime slot comprised the parts of the day when most Americans (men) were assumed to be at work, leaving their wives at home to take care of the domestic chores. Soap operas, then, worked under the assumption that housewives were their main audience; their creators tried to appeal to those housewives and the companies that wanted to advertise to them through cheaply made programming sponsored by soap companies. The prime-time time slot, on the other hand, was the time slot with the biggest ad revenue and was usually defined as 8:00 pm (after the assumed dinner hour) to midnight.[18]

Daytime soap operas were built on expectations about the intended audience and "the rhythms of women's work in the home."[19] Appealing to housewives informed the format and structure of soap operas. For instance, the scene repetition before and after commercial breaks and frequent repetition of key information allowed women to follow the narrative while completing their work in the home during an era in which television could not be recorded or "rewound." She would not have to worry about missing a plot beat if her work called her away from the television screen; she could be confident it would be repeated. This underpinned the development of narrative predictability in soap operas. This narrative predictability is considered fundamental to the genre; shows without it are not generally considered "soap operas," as opposed to just serial melodramas.

Dallas set off scholarly debate about the boundaries of the genre, with scholars staking out firm positions. Writing in 1980, Communications and soap opera scholar Mary Cassata argued, "I do not agree with the majority opinion in defining the current prime-time favorite, *Dallas,* as a soap opera, (it plays too fast and loose with the predictableness of soap opera conventions)"[20] Cassata argues that predictableness—or narrative stability—of a show was for some scholars in defining what was a soap opera. In other words, people expected a viewer to be able to predict plot beats. This relates back to the format of soap operas being influenced by the domestic workflow, as viewers would not have to pay close attention while watching to identify character archetypes, for example.

The low expectations built into the time slot influenced perceptions of the value of daytime programming, and soap operas have been long devalued by scholars and the wider audience alike. Soap opera scholar Robert C. Allen confronted the issue of soap operas not being taken seriously by critics in his introduction to T*o Be Continued.... Soap Operas Around the World*. He identified the contradiction of their popularity among audiences as opposed to critical disdain and notes that it is "primarily a function of the status of soap operas as a gendered form of narrative and its resistance to being read according to the protocols of more closed narrative forms."[21] He argues that soap operas, as serial melodramas, tend to have a "resistance to closure." In other words, they do not tell a story with an endpoint in mind; moreover, that this convention is specific to programming defined as appealing to women, further devalues the narrative form.[22] For critics, soap operas' resistance to closure made it harder to analyze television in the same traditional ways that film has been studied and evaluated. Thus, not having a clear end point made it easier for those coming in and out of the room to know what is happening, but also for critics to dismiss soap operas as 'bad' television.

Soap operas, then, have a bad reputation, and have even been disregarded as "trash." The trashy connotation of melodramas and soap operas runs deep, and is connected with their perceived address to women viewers. This association is literal as well as metaphorical: Allen notes that soap operas have been connected to 'dirt' in many ways, characters collecting dirt on each other to expose and gossip about, and literal dirt that needs to be cleaned with products that are conveniently advertised to the viewers.[23] This dirt was transferred to their fans, who were not looked highly upon. As Allen notes, soap opera audiences were considered a "distinctively different audience group with special needs and lacks," that was uniquely gendered as feminine.[24] These "special needs and lacks" of course, differentiated women from the male audience, perceived as more sophisticated and discriminating, and contemporaries assumed men would not be interested in, much less watch, soap operas. Women internalized this critique. Writing in 1995, Harrington and Bielby noted that "people are embarrassed by their own viewing habits . . . and routinely lie about how much television they watch or which programs they prefer."[25] This was particularly the case before the current era of "platinum television," when there was much public (and especially intellectual) derision of television.[26]

Soap operas, appealing to those particular 'dirty' feminine needs and special feminine audiences, also centered women's stories more than their prime time counterparts. In 1980 Michelle Lynn Rondina, Mary Cassata, and Thomas Skill analyzed the lifestyles and demographics of daytime television's characters. They found that daytime shows had 49.2 percent women characters while prime time shows only had 29 percent.[27] Not only did television shows that aired in the daytime slot present more women characters to explore—characters for their presumed female audience to relate to—but these women also exercised more agency than women in primetime.[28] Consider the examples of the mostly male, action-oriented shows on prime-time in the 1980s, such as Magnum P.I (see Joshua

Buckingham's contribution to this volume). Advertisers and network executives thought stories that centered more women would lose their money-making audience in prime time. This was important because soap operas were also often cheaply made; they were not expensive investments with high rewards. They were cheaply produced for "the masses."[29]

Since soap operas were the mass produced "trash" that was only put out because it filled time and made money, they were not usually scheduled in prime time. Prime time was the prestigious big money time slot for 'legitimate' television. But, it was undeniably a popular genre, and there was precedent for it: for instance, ABC's soap *Peyton Place* aired during prime-time in the 1960s, demonstrating that soap opera genre conventions of a repetitive and predictable story structure with an open narrative, and a focus on women's stories could find an audience in prime time. In the late 1970s, *Dallas* defied expectations, introducing soap opera narrative to prime-time with blockbuster results.

DALLAS

Dallas, a prime-time serial melodrama created by David Jacobs, aired on CBS from 1978 to 1991.[30] *Dallas* follows three generations of the Ewing clan, a wealthy Texan oil family, headed by patriarch Jock (Jim Davis) and his wife Ellie (Barbara Bel Geddes). The show mostly focuses on the second generation of Ewings, Jock and Ellie's sons—the respectable modern man Bobby (Patrick Duffy) and his scheming brother J.R. (Larry Hagman)—and their wives, the precocious Pamela (Victoria Principal), and the emotionally fraught and complicated Sue Ellen (Linda Gray) respectively. They have a third son, Gary, but he is estranged from the family and appears less often. Gary's troubled teen daughter Lucy (Charlene Tilton), who lives separately from her father on the Southfork ranch with the rest of the family, is also a regular character.

As a prime-time serial melodrama, *Dallas* had already broken one of the genre conventions of soap opera by not airing in the daytime. Despite this and bending other conventions it was the "only current prime-time soap opera," for a time.[31] *Dallas*' commitment to melodrama made it believable as a soap opera, which David Jacobs noted himself in a quote from *The Chicago Tribune*: Jacobs said that he was not offended when *Dallas* was called a soap opera because it "refers only to the serialized and melodramatic elements."[32] This shows how, for many people, melodrama was the most important genre convention of a soap opera, and *Dallas* had it in spades.

Dallas did not air during daytime, but it did have a repetitive story with an open narrative. The show has some plot points that were hit in every episode, most notably J.R's scheme of the week. *Dallas* also never had an endpoint in mind; even after the final season in 1991 there was not a true end. Indeed, in 2012 *Dallas* was "rebooted" and had many of its original actors return, including Patrick Duffy (Bobby) and Larry Hagman (J.R.).

While *Dallas* did have repetitive plot beats, the narrative was never stable. *Dallas*' commitment to trying out different plots did not always work out in its favor. A famous example of this is the ninth season of *Dallas*. The show killed Bobby in the season eight cliffhanger, only to bring him back in the season nine finale. The season nine cliffhanger revealed that the whole season had been all Pamela's dream and Bobby was still alive. Retconning the entire season was very unpopular with fans and critics alike because it was seen as a cheap way to get Patrick Duffy back in the show. One magazine columnist wrote that he had "caught three flies so far," from his mouth hanging open after seeing how Bobby had 'survived' being run over by a car.[33]

Another example of *Dallas'* narrative instability came much earlier in the show's run and was much more popular with fans. The "Who shot J.R.?" plotline was advertised so much that even people who were not invested in *Dallas* were aware of it—and quite annoyed to be confronted with it all summer long. For example, one newspaper noted that "the scene had been repeated on television at least a dozen times this week in commercials designed to keep us guessing about whodunit."[34](Spoiler: it was his sister-in-law—Kristin Shepard played by Mary Crosby—with whom he was having an affair.) Who shot J.R.? was also representative of a shift in focus on *Dallas*. Originally, the show focused more on Pamela trying to fit into the Ewing clan, but Larry Hagman's performance as J.R. took audiences by storm, and the show focused on him more and more.

Who shot J.R. is an example of another genre convention that *Dallas* bent: the soap opera's usual focus on women's stories. *Dallas* does have women characters who exercise agency and make choices. For example, teenager Lucy especially likes to make her own choices, like skipping school, usually to the annoyance of the adults in her life. However, men's stories undeniably take precedent. This is not uncommon for prime-time serial melodramas; one scholar in 1983 even noted that "the primary difference between prime time programs such as *Dallas* and *Dynasty* and the daytime serials is the centrality of a male character to the narrative."[35]

Dallas centered its male characters in ways that were less common in daytime soap operas. Focusing on men was not rare; scholars have long recognized the centrality of men and men's stories in film and television. Not only did prime time programs center male characters and stories, but their depiction of women privileged men's reception. This phenomenon, notably described by film theorist Laura Mulvey, is called the "male gaze." Mulvey articulated the idea of the male gaze in the 1970s to describe how and why stories in film center men both as characters and as viewers. In her widely-read article "Visual Pleasure and Narrative Cinema," she discusses a psychoanalysis of "scopophilia," or the pleasure audiences find in seeing. She argues that in film, women are symbols that do not serve film narrative, but are passive objects to be consumed by the spectator, an image that stops the flow of the story to look at them in pleasure.[36] Mulvey defines scopophilia as "pleasure in looking at another person as an erotic object"[37] and argues that scopophilia, alongside "ego libido" (which she identifies as "identification processes ") are the structures that create the male gaze in a patriarchal society.[38]

Mulvey only discusses the male gaze in relation to cinema, but I would argue that the male gaze is relevant to television also. This is because television is also created in a patriarchal society and also has moments where women are objects that are looked at only for visual pleasure, especially in 'prestige' prime time television shows like *Dallas*. The Ewing men are the center of the family, and their wives (categorically subordinate because they are not blood Ewings) serve the show primarily as their husband's relations, even when acting on their own.

On *Dallas*, even plot lines that ostensibly center women characters always take a moment to focus on the men's emotions and reactions. In season one episode two of *Dallas,* titled "Lessons," a major plot point focuses on the relationship between two women, Pamela and Lucy. Lucy has been missing school due to reasons unknown to the Ewing men. Pam tries to take on a maternal role to guide Lucy back to the right path and also solidify her own place in the Ewing family. For these two characters, connected only through their respective relationships to Bobby, this premise introduces the opportunity to establish a relationship independent of men; it also allows the show to "spend" airtime exploring the nature of women's relationships and the difficulties they face as new wives and young adults. Instead, the plot develops to further subordinate these characters to their male counterparts.

On a date, Pamela and Bobby go to a disco, where they find out that Lucy is having an affair with their much older ranch hand Ray Krebbs (Steve Kanaly). Pamela relegates her own struggles to take on a maternal role and to try to be a good wife for Bobby. The discovery of the affair is important because it reveals to Pamela and Bobby why Lucy was skipping school, but it also introduces a love triangle among Ray, Bobby, and Pamela. Ray is Pamela's ex-boyfriend and her current husband's employee. Rather than allowing Pam to build a relationship with Lucy independent of *Dallas*' men, this plot focuses on Lucy and Ray's affair and the family's—especially Pam's—reaction to it. It sets up Pamela as a sexual object: she had a premarital relationship with a man who is not her husband, which potentially destabilizes her relationship with Bobby while also defining her as "guilty" with something to prove. Lucy and Pamela spend a good few minutes alone, but their conversations are haunted by Ray Krebbs' presence. They argue, in part because Lucy knows that Ray and Pamela used to be a couple. Ultimately, the show displaces the Pamela/Lucy relationship with the Bobby/Ray relationship. After he finds out about the affair, Bobby punches Ray, fulfilling his role as patriarch, protecting while also establishing paternalistic control over his niece.

Alongside Pam, Lucy—a teenager—is portrayed as sexually available to the men of the Ewing universe and, by extension, the (male) audience. At the disco, the show stops the action for a moment to focus on Pamela dancing. The camera takes on Bobby's point of view lingering on her backside as she dances. Pamela is not *only* a passive visual object as described in Mulvey's article, but instances like this in "Lessons" make it clear that there are times that she serves the show as a passive visual object, and it can take precedence over plot elements. It is also important to note that while Pamela has some agency in this story line, her actions are motivated by wanting to be a better wife to Bobby by being accepted by the Ewings. I would argue that the camera lingering on her and the camera's identification with Bobby's perspective, constitutes the male gaze, as defined by Mulvey. It also informs how this plot point played out: it became more about Pamela being set up as a good wife and (surrogate) mother (and a good addition to the Ewing clan) rather than Pamela being an active agent.

Setting up women to be good wives to the Ewing men was central to the representation of family and relationships between men and women in *Dallas* and relates to the Reagan-era family values that the *Dallas* portrayed. For instance, it is eventually revealed that working-class outsider Ray Krebbs is not only the rancher in charge of the cattle at Southfork Ranch, but also Jock's illegitimate child. By implication, the audience now knows that Jock cheated on Miss Ellie and that Lucy had an affair with her uncle.[39] Despite his claim to the Ewing name and fortune, Ray remains loyal to the family and keeps his secret so as not to destroy the family and Miss Ellie in particular. But such secrets cannot be kept forever, and, when Ellie does find out, she forgives Jock for cheating on her. The show deals with this betrayal in the span of just one episode, and it sets her up as the forgiving southern matriarch and the perfect companion to Jock's southern patriarch. The southern part of this is important as part of the show's melodrama. The south has a history of being portrayed as a part of the country that is loyal to traditional values and glorifies the past. Consider, for example, the enduring myth of *Gone with the Wind*. Miss Ellie is depicted as a 'good wife' and mother, similar to how "Lessons" set up Pamela as a good wife and future mother. This comes at the cost of focusing on women's stories, motives, dreams, and relationships, since the stories are no longer about how women relate to each other, but instead how they can be good for their family.

Dallas focuses on a business family instead of a community. The conflicts are familial and business-related and cannot easily be separated, especially after Ray is revealed to be Jock's son and Pamela marries into the family. The family is fairly insular and outsiders are shown to either be inconsequential or dangerous. Interloping women are sexual objects usually having an affair with J.R., while men pose physical threats to the family. For instance, in season one episode four "Winds of Vengeance," the sins of the sons are revisited upon the family. After J.R. and Ray commit adultery with two women, two of their male relatives find out where the Ewings live and go to Southfork to take revenge.

The two men, who are portrayed as working-class (and specifically do not harm Pamela because she also comes from a working-class background) take the family hostage. They rough up Sue Ellen and Lucy, left defenseless by their "protector's" indiscretions, and prepare to rape them. The incident lays bare the boundaries of the *Dallas* world. Men can act out indiscriminately or take revenge; women are at the mercy of the men's decisions; working-class people are unnecessarily cruel and uncouth, not to be trusted, unless they try to fit into the world of the family like Pamela has; and wealthy men can "get away with it." Jock and Bobby make it back to Southfork in time to stop any further violence, but not before the outsiders make it clear that only family is trustworthy.

Dallas is not unique in focusing on a wealthy family over a community in prime time serial melodramas. Ellen Seiter has noted that in many prime time melodramas "the plot focuses entirely on an upper class family, their spouses, business rivals, or illegitimate relations. Working-class characters are almost entirely absent."[40] *Dallas* and its imitator *Dynasty* are prime examples of that. Moreover, the focus on wealthy characters and rich settings is part of the materialism that was common in art created in the 1980s (see Aaidin Cuinfield's contribution to this volume), and places *Dallas* right in the middle of common media tropes of the time.

The setting of *Dallas*, Southfork, is a large, working ranch in the suburbs of the Texas city, Dallas. This setting is a part of the nostalgic, materialistic appeal of the show. The Ewings have the wealth from oil but are still clinging to the ruralness of the ranch. This creates *Dallas*' own "cowboy dreams" that Le Coney and Trodd wrote about.[41] This nostalgic setting did not go unnoticed by contemporary viewers. Observers often compared *Dallas* to James Dean's last movie *Giant* (1956); certainly they both follow Texan oil families with rivalries over long periods of time.[42] There are many parallels in the plots of both *Giant* and *Dallas*, as for example the common plot point of two kids from the rival families falling in love. This Romeo and Juliet trope also represented a point of nostalgia for the 'Fifties: *West Side Story* (1961) similarly adapted Shakespeare's *Romeo and Juliet* and was released early enough in the 1960s to be lumped in with the cultural idea of the 'Fifties.

While *Dallas* fits right into the nostalgic concept of "the Fifties" that Le Coney and Trodd wrote about, it is not just pure 'Fifties nostalgia; instead it is all about the 1980s. A patriarchal family, wealthy, and living on a working Texan ranch while also owning and operating a successful oil business is a fantastic example of cowboy mythology mixed with the materialism and "family values" espoused by the Reagan Administration in the 1980s. The modernization that came with the 1980s is especially noticeable in women's roles. For instance, Pamela gets to work a job, but it is significant that it was not an executive job like Bobby and J.R. Instead, she works at a clothing shop, coded as "temporary" (until she starts having a family), while also giving the show an avenue to introduce 1980s fashion. In addition, Lucy is allowed to have a lot of premarital sex without being killed off or punished in a permanent way. This allows for audiences to enjoy *Dallas* through not only scopophilia, but also through imagining a world with modern comforts, like discos and fancy new cars and technology, that still has the family values of 'the Fifties,' as discussed in Le Coney and Trodd's article.

This all relates to what people found pleasurable to watch in *Dallas*. Many scholars have looked into what makes television fun to watch, and *Dallas* has been a part of that discussion for a long time. Cultural studies scholar Ien Ang wrote her first book in 1985 entirely on the subject of what people found pleasurable about *Dallas*. *Watching Dallas* looks at forty-two letters Dutch *Dallas* fans wrote in response to a prompt that asked them to describe why they dis/like/d watching *Dallas*.[43] One theme that came up in the letters she analyzed was escapism. One letter Ang analyzed from a fan noted how the glitz and glamor of *Dallas'* sets and characters were part of the appeal, the letter said, "First of all it's entertainment for me, part show, expensive clothes, beautiful horses."[44] The focus on this causes Ang to conclude that there is, for some viewers, a "'flight' into a fictional fantasy world." Ang notes, however, that this is not a denial of reality, instead viewers are "playing with it."[45] Playing with reality is the fun aspect of melodrama, which is more concerned with emotional realism than actual realism. In this the fans that Ang analyzed acknowledged that *Dallas* is a fun, entertaining world to imagine yourself in and play around with.

Among American viewers, *Dallas* invoked the "fantasy Fifties," where nuclear families with patriarchs and white picket fences were normal and uncritically acclaimed as good. While a Dutch audience did not have the same cultural context for the 'the Fifties' as the American audience, Ang's analysis offers some other points of pleasure people found while watching *Dallas,* notably perceived genuineness, or 'realness.'[46] The plot devices of *Dallas* may not seem very realistic but Ang argues that what matters to audiences is that characters *feel* emotionally real.[47]

Pamela's experiences in "Lessons" would fit right into this reading. Although the specific details might seem extreme—sexual relations between a young woman and a much older man, the exploitative nature of which is pointed up by the family's concern for her missed school days—the fundamental motivations are recognizable to viewers, both from their own lives and the persistence of certain cultural tropes. Meeting and bonding with your new in-laws, love triangles, and adultery are realistic and recognizable tropes. *Dallas* took the reality of fitting in with the in-laws and made it fantastical and glamorous by adding in the setting of the disco, and having it be acted out by the beautiful actors who play the Ewings.

This focus on the wealthy Ewings and their problems was appealing in the 1980s, then, because of the real political, economic, and social situations Americans found themselves in. Perhaps paradoxically, *Dallas* arrived on the airwaves in the midst of an economic recession; that is, a show that glorified great wealth and materialistic consumerism found great popularity at a time when many Americans were suffering financially. Some observers argued that precisely this paradox contributed to *Dallas*' appeal. In a 2012 interview Larry Hagman (J.R.) recalled that "'... *Dallas* was popular during a recession, the Reaganist recession. And people couldn't afford to go out and get a babysitter, go out and get dinner, have dinner and go to a movie.'"[48] Similarly, *Dallas*' producer Leonard Katzman said in an article in *The Chicago Tribune* that "its appeal is voyeurism:" people wanted to imagine themselves as rich as the Ewings during the recession, and the producers, showrunners, and actors knew it.[49]

CONCLUSION

Dallas bent soap opera genre conventions enough to be 'respectable' to a mainstream prime-time audience. It drew upon the appeal of soap opera conventions like melodrama and open narrative and bent them by airing on prime-time, focusing on male characters, and having conflicts that primarily were business-related instead of relational. *Dallas* aired during the 1980s when the United States had multiple national stresses because of the economic crisis of stagflation and a perceived rise in crime, which made the cultural idea of 'the fifties' appealing to audiences and voters because it represented a conservative shift in the political and social climate of the United States in the 1980s. *Dallas* glorified this 'fantasy fifties' through its Texas setting at Southfork Ranch and its patriarchal family structure, while being deeply entrenched in the values of the present.

Dallas was so nationally and internationally popular that it transformed the way that both Americans and other people in the world thought about the United States. *Dallas* appealed to so many people because it was a fun, melodramatic show that played with long-standing tropes and genre conventions to tell a story about a wealthy, beautiful business family. The melodramatic nature of *Dallas* also contributed to how people consumed it. The moral legibility of the show was strong, and it offered a vision of the future for America, similarly to Reagan, that called back to 'fifties family values: patriarchal family structures like the nuclear family, the downplaying of working-class people, and the romanticization of the rural west but with the exciting new twist of materialism. This was all shown through the stories *Dallas* chose to tell, like Pamela's story in "Lessons," and through how those stories were told, notably the male gaze as described by Laura Mulvey.

Notes

1. Robert P. Weiss, "Privatizing the Leviathan State: A 'Reaganomic' Legacy," in *The 1980s: A Critical and Transitional Decade*, edited by Kimberly R. Moffitt and Duncan A. Campbell, (New York, NY: Lexington Books, 2011), 89-107.
2. In the 1970s the country began to deal with stagflation. Stagflation is the phenomena of "simultaneously high unemployment and inflation," and it was dampening the American economy. The cause of stagflation had many factors, one of which was the collapse of the Bretton Woods economic system. The Bretton Woods system was a system that was supposed to regulate the trade of gold. The system collapsed in part because it was weakened by Nixon separating the dollar from the value of gold, which was meant to stop inflation but instead increased it, and the economic consequences of the disastrous war in Vietnam. That entire situation was compounded by the Arab oil embargo in 1973, which shocked American industry because of the fuel crisis it caused. Editors of the Encyclopedia Britannica, "Decline of Gold," in *Encyclopedia Britannica*, accessed May 2, 2023, https://www.britannica.com/topic/political-economy/Historical-development#ref255576; and Editors of the Encyclopedia Britannica, "Money," in *Encyclopedia Britannica*, accessed May 2, 2023, https://www.britannica.com/topic/money/The-decline-of-gold#ref1089594.
3. Robert P Weiss, "Privatizing the Leviathan State," 90-91.
4. Alan Bitton, "Matter and Mammon: Fiction in the Age of Reagan," *The 1980s: A Critical and Transitional Decade,* edited by Kimberly R. Moffitt and Duncan A. Campbell, (New York, NY: Lexington Books, 2011), 417-435.
5. Michael Dwyer, "'Fixing' the Fifties: Alex P. Keaton and Marty McFly," in *The 1980s: A Critical and Transitional Decade*, edited by Kimberly R. Moffitt and Duncan A. Campbell (New York, NY: Lexington Books, 2011), 201-223, 202.
6. Michael Dwyer, "'Fixing' the Fifties," 202.
7. Matt Seitz, "Why the Golden Age of TV Was Really Born in the 1980s" *Vulture*, accessed May 1, 2023, https://www.vulture.com/2016/10/golden-age-of-tv-was-born-in-the-1980s.html.
8. Christopher Le Coney and Zoe Trodd, "Reagan's Rainbow Rodeos: Queer Challenges to the Cowboy Dreams of Eighties America," *Canadian Review of American Studies*, Vol. 39, No. 2 (2009): 163-183; 1.
9. Le Coney et al, "Reagan's Rainbow Rodeos," 164.
10. I look mostly at the early seasons of *Dallas*, only analyzing a few plot points from later seasons. I did this because *Dallas*' popularity took off in the early seasons and it narrows the scope of this paper.
11. Michael Stewart, "Introduction: Film and TV Melodrama: An Interview," in *Melodrama in Contemporary Film and Television*, (UK: Palgrave MacMillan: 2014), 1-26; 3. See Stewart for a literature review of scholarship on melodrama.
12. Daniel Chandler and Rod Munday "Melodrama," in *A Dictionary of Media and*

Communication, 3rd ed, (Oxford University Press: 2020), accessed May 1, 2023, https://www-oxfordreference-com.ezproxy.lib.vt.edu/display/10.1093/acref/9780198841838.001.0001/acref-9780198841838-e-1705?rskey=YdWciX&result=1.

13. Michael Stewart, "Introduction," 3.
14. Michael Stewart, "Introduction," 1.
15. Michael Stewart, "Introduction," 2.
16. Robert C. Allen, "Introduction," in *To Be Continued*, edited by Robert C. Allen, (New York, NY: Routledge, 1995), 1-2.
17. Robert C. Allen, "Introduction," 3-4.
18. Daniel Chandler and Rod Munday, "Prime Time," in *A Dictionary of Media and Communication*, 3rd ed, (Oxford University Press: 2020), accessed May 1, 2023, https://www-oxfordreference-com.ezproxy.lib.vt.edu/display/10.1093/acref/9780198841838.001.0001/acref-9780198841838-e-2138?rskey=6gLoYj&result=1.
19. C. Lee Harrington and Denise D. Bielby, "Chapter One: The Soap Opera Fan World," in *Soap Fans: Pursuing Pleasure and Making Meaning in Everyday Life*, (Philadelphia, PA: Temple University Press, 1995), 17.
20. Mary Cassata, "The More Things Change the More they are the Same: An Analysis of Soap Operas from Radio to Television," in *Life on Daytime Television: Tuning-In American Serial Drama*, (Norwood, NJ: Ablex Publishing Corporation, 1983), 99.
21. Robert C. Allen, "Introduction," 3.
22. Martha Nochimson, "Epilogue: What is Normal," in *No End to Her: Soap Opera and the Female Subject* (Los Angeles, CA: University of California Press, 1992), 194.
23. Robert Allen, "Introduction," 4.
24. Robert Allen, Introduction, 6.
25. C. Lee Harrington and Denise D. Bielby, "Introduction," 4.
26. David Bianculli, *The Platinum Age of Television : From I Love Lucy to the Walking Dead, How Tv Became Terrific* (New York: Doubleday, 2016).
27. Michelle Lynn Rondina, Mary Cassata, and Thomas Skill, "Placing a 'Lid' on Television Serial Drama: An Analysis of the Lifestyles, Interpersonal Management Skills, and Demography of Daytime's Fictional Population," in *Life on Daytime Television: Tuning-In American Serial Drama*, (Norwood, NJ: Ablex Publishing Corporation, 1983), 9.
28. In the 1950s Lucille Ball's role in *I Love Lucy* gave a lot of importance to Lucy's relationship with her best friend Ethel. Ball also had a level of artistic control over the show that was similar to film directors. In the late 1960s Mary Tyler Moore started her production company MTM, which was quickly followed by the *Mary Tyler Moore Show*. MTM was prolific in the following decades and had its hand in many influential sitcoms, like *The Bob Newhart Show* and *Cheers*.

Editors of Encyclopedia Britannica, "Mary Tyler Moore," *Encyclopedia Britannica*, accessed May 1, 2023, https://www.britannica.com/biography/Mary-Tyler-Moore.

29. Robert C. Allen, "Introduction," 4.

30. "*Dallas*," *IMDB*, accessed May 1, 2023. https://www.imdb.com/title/tt0077000/.

31. Joel Swerdlow, "Tempo: '*Dallas*' 'Peyton Place' all Over again—and a Prime-Time Success '*Dallas*': Prime-Time Soap Opera Becomes a Legend in its Own Time Slot," *Chicago Tribune* (1963-1996), Jan. 8, 1980. http://login.ezproxy.lib.vt.edu/login?url=https://www.proquest.com/historical-newspapers/tempo/docview/1 70107100/se-2.

32. This quote also suggests Jacobs' attempt to minimize *Dallas*' association with that lower art form. Swerdlow, "'Dallas' 'Peyton Place' all Over again."

33. Marvin Kitman, "Pam Ewing's Bad Dream," *New Leader*, Oct. 6, 1986, 21. http://login.ezproxy.lib.vt.edu/login?url=https://www.proquest.com/magazines/pam-ewings-bad-dream/docview/1308973498/se-2.

34. Leah Y. Latimer. "Hey, they Shot J.R.!: 'Dallas' Villain Plugged; it's a Contract Job 'Hey, they Shot J.R.!'" *The Washington Post,* Mar. 22, 1980. http://login.ezproxy.lib.vt.edu/login?url=https://www.proquest.com/historical-newspapers/hey-they-shot-j-r/docview/147234866/se-2.

35. Ellen Seiter, "Men, Sex and Money in Recent Family Melodramas," *Journal of the University Film and Video Association*, Vol. 35, no. 1 (1983): 17–27. http://www.jstor.org/stable/20686924.

36. Laura Mulvey, "Visual Pleasure and Narrative Cinema," in *Film Theory and Criticism: Introductory Readings*, fifth ed, edited by Leo Braudy and Marshall Cohen, (New York, NY: Oxford University Press, 1999), 801-816; 809.

37. Mulvey, "Visual Pleasure," 815.

38. Mulvey, "Visual Pleasure," 815.

39. *Dallas*, Season 4 episode 7, "The Fourth Son," directed by Irving J. Moore, written by David Jacobs, aired on CBS, December 12, 1980.

40. Ellen Seiter, "Men, Sex and Money in Recent Family Melodramas," 18. http://www.jstor.org/stable/20686924.

41. Le Coney et al, "Reagan's Rainbow Rodeos," 163-183.

42. Peter Applebome, "What Dallas Thinks of 'Dallas': What they Think of 'Dallas' in Dallas," *New York Times*, Jan. 9, 1980. http://login.ezproxy.lib.vt.edu/login?url=https://www.proquest.com/historical-newspapers/what-dallas-thinks/docview/121017200/se-2.

43. Ien Ang, *Watching Dallas: Soap Opera and the Melodramatic Imagination*, trans by Della Couling, (New York, NY: Methuen, 1982), 10.

44. Ang, *Watching Dallas*, 25.

45. Ien Ang, *Watching Dallas*, 49.

46. Ien Ang, *Watching Dallas*, 30.

47. Ien Ang, Watching *Dallas*, 47. This suggests also that *Dallas* is a melodrama in accordance with Stewart's definition of melodrama.

48. "'Dallas' Returns With Original and New Cast," *Associated Press,* June 12 2012, accessed May 1, 2023. https://fod-infobase-com.ezproxy.lib.vt.edu/p_ViewVideo.aspx?contentID=VZbz1-SHPYM&channel=Associated%20Press&chnID=57.

49. Swerdlow, "'Dallas' 'Peyton Place' all Over again.

Made in United States
Troutdale, OR
11/03/2026

56844893R00139